ROOT CAUSE FAILURE ANALYSIS

PLANT ENGINEERING MAINTENANCE SERIES

Vibration Fundamentals
R. Keith Mobley

Root Cause Failure Analysis
R. Keith Mobley

Maintenance Fundamentals
R. Keith Mobley

ROOT CAUSE FAILURE ANALYSIS

R. Keith Mobley

Newnes

Boston   Oxford   Auckland   Johannesburg   Melbourne   New Delhi

Library of Congress Cataloging-in-Publication Data

Mobley, R. Keith, 1943-
    Root cause failure analysis / by R. Keith Mobley.
      p. cm. — (Plant engineering maintenance series)
    Includes index.
    ISBN 0-7506-7158-0 (alk. paper)
    1. Plant maintenance.   2. System failures (Engineering)
I. Title. II. Series.
TS192.M625 1999
658.2'02—dc21   98-32097
                                                 CIP

British Library Cataloguing-in-Publication Data
A catalogue record for this book is available from the British Library.

The publisher offers special discounts on bulk orders of this book.
For information, please contact:
Manager of Special Sales
Butterworth–Heinemann
225 Wildwood Avenue
Woburn, MA 01801–2041
Tel: 781-904-2500
Fax: 781-904-2620

For information on all Newnes publications available, contact our World Wide Web home page at: http://www.newnespress.com

10 9 8 7 6 5 4 3 2 1

Printed in the United States of America

CONTENTS

Part I   Introduction to Root Cause Failure Analysis   1

    Chapter 1   Introduction   3
    Chapter 2   General Analysis Techniques   6
    Chapter 3   Root Cause Failure Analysis Methodology   14
    Chapter 4   Safety-Related Issues   58
    Chapter 5   Regulatory Compliance Issues   62
    Chapter 6   Process Performance   72

Part II   Equipment Design Evaluation Guide   75

    Chapter 7   Pumps   77
    Chapter 8   Fans, Blowers, and Fluidizers   97
    Chapter 9   Conveyors   112
    Chapter 10   Compressors   123
    Chapter 11   Mixers and Agitators   147
    Chapter 12   Dust Collectors   153
    Chapter 13   Process Rolls   164
    Chapter 14   Gearboxes/Reducers   171
    Chapter 15   Steam Traps   187
    Chapter 16   Inverters   194
    Chapter 17   Control Valves   202
    Chapter 18   Seals and Packing   220

Contents

Part III   Equipment Troubleshooting Guide   237

Chapter 19   Pumps   239
Chapter 20   Fans, Blowers, and Fluidizers   246
Chapter 21   Conveyors   251
Chapter 22   Compressors   254
Chapter 23   Mixers and Agitators   264
Chapter 24   Dust Collectors   266
Chapter 25   Process Rolls   269
Chapter 26   Gearboxes or Reducers   271
Chapter 27   Steam Traps   276
Chapter 28   Inverters   278
Chapter 29   Control Valves   280
Chapter 30   Seals and Packing   282
Chapter 31   Others   285

List of Abbreviations   288
Glossary   291
References   305
Index   306

Part I

INTRODUCTION TO ROOT CAUSE FAILURE ANALYSIS

1

INTRODUCTION

Reliability engineering and predictive maintenance have two major objectives: preventing catastrophic failures of critical plant production systems and avoiding deviations from acceptable performance levels that result in personal injury, environmental impact, capacity loss, or poor product quality. Unfortunately, these events will occur no matter how effective the reliability program. Therefore, a viable program also must include a process for fully understanding and correcting the root causes that lead to events having an impact on plant performance.

This book provides a logical approach to problem resolution. The method can be used to accurately define deviations from acceptable performance levels, isolate the root causes of equipment failures, and develop cost-effective corrective actions that prevent recurrence. This three-part set is a practical, step-by-step guide for evaluating most recurring and serious incidents that may occur in a chemical plant.

Part One, Introduction to Root Cause Failure Analysis, presents analysis techniques used to investigate and resolve reliability-related problems. It provides the basic methodology for conducting a root cause failure analysis (RCFA). The procedures defined in this section should be followed for all investigations.

Part Two provides specific design, installation, and operating parameters for particular types of plant equipment. This information is mandatory for all equipment-related problems, and it is extremely useful for other events as well. Since many of the chronic problems that occur in process plants are directly or indirectly influenced by the operating dynamics of machinery and systems, this part provides invaluable guidelines for each type of analysis.

Part Three is a troubleshooting guide for most of the machine types found in a chemical plant. This part includes quick-reference tables that define the common failure or deviation modes. These tables list the common symptoms of machine and process-related problems and identify the probable cause(s).

PURPOSE OF THE ANALYSIS

The purpose of RCFA is to resolve problems that affect plant performance. *It should not be an attempt to fix blame for the incident.* This must be clearly understood by the investigating team and those involved in the process.

Understanding that the investigation is not an attempt to fix blame is important for two reasons. First, the investigating team must understand that the real benefit of this analytical methodology is plant improvement. Second, those involved in the incident generally will adopt a self-preservation attitude and assume that the investigation is intended to find and punish the person or persons responsible for the incident. Therefore, it is important for the investigators to allay this fear and replace it with the positive team effort required to resolve the problem.

EFFECTIVE USE OF THE ANALYSIS

Effective use of RCFA requires discipline and consistency. Each investigation must be thorough and each of the steps defined in this manual must be followed.

Perhaps the most difficult part of the analysis is separating fact from fiction. Human nature dictates that everyone involved in an event or incident that requires a RCFA is conditioned by his or her experience. The natural tendency of those involved is to filter input data based on this conditioning. This includes the investigator. However, often such preconceived ideas and perceptions destroy the effectiveness of RCFA.

It is important for the investigator or investigating team to put aside its perceptions, base the analysis on pure fact, and not assume anything. Any assumptions that enter the analysis process through interviews and other data-gathering processes should be clearly stated. Assumptions that cannot be confirmed or proven must be discarded.

PERSONNEL REQUIREMENTS

The personnel required to properly evaluate an event using RCFA can be quite substantial. Therefore, this analysis should be limited to cases that truly justify the expenditure. Many of the costs of performing an investigation and acting on its recommendations are hidden but nonetheless are real. Even a simple analysis requires an investigator assigned to the project until it is resolved. In addition, the analysis requires the involvement of all plant personnel directly or indirectly involved in the incident. The investigator generally must conduct numerous interviews. In addition, many documents must be gathered and reviewed to extract the relevant information.

In more complex investigations, a team of investigators is needed. As the scope and complexity increase, so do the costs.

As a result of the extensive personnel requirements, general use of this technique should be avoided. Its use should be limited to those incidents or events that have a measurable negative impact on plant performance, personnel safety, or regulatory compliance.

WHEN TO USE THE METHOD

The use of RCFA should be carefully scrutinized before undertaking a full investigation because of the high cost associated with performing such an in-depth analysis. The method involves performing an initial investigation to classify and define the problem. Once this is completed, a full analysis should be considered only if the event can be fully classified and defined, and it appears that a cost-effective solution can be found.

Analysis generally is not performed on problems that are found to be random, nonrecurring events. Problems that often justify the use of the method include equipment, machinery, or systems failures; operating performance deviations; economic performance issues; safety; and regulatory compliance issues.

2

GENERAL ANALYSIS TECHNIQUES

A number of general techniques are useful for problem solving. While many common, or overlapping, methodologies are associated with these techniques, there also are differences. This chapter provides a brief overview of the more common methods used to perform an RCFA.

FAILURE MODE AND EFFECTS ANALYSIS

A failure mode and effects analysis (FMEA) is a design-evaluation procedure used to identify potential failure modes and determine the effect of each on system performance. This procedure formally documents standard practice, generates a historical record, and serves as a basis for future improvements. The FMEA procedure is a sequence of logical steps, starting with the analysis of lower-level subsystems or components. Figure 2–1 illustrates a typical logic tree that results with a FMEA.

The analysis assumes a failure point of view and identifies potential modes of failure along with their failure mechanism. The effect of each failure mode then is traced up to the system level. Each failure mode and resulting effect is assigned a criticality rating, based on the probability of occurrence, its severity, and its delectability. For failures scoring high on the criticality rating, design changes to reduce it are recommended.

Following this procedure provides a more reliable design. Also such correct use of the FMEA process results in two major improvements: (1) improved reliability by anticipating problems and instituting corrections prior to producing product and (2) improved validity of the analytical method, which results from strict documentation of the rationale for every step in the decision-making process.

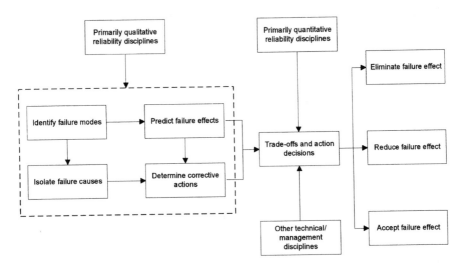

Figure 2–1 Failure mode and effects analysis (FMEA) flow diagram.

Two major limitations restrict the use of FMEA: (1) logic trees used for this type of analysis are based on probability of failure at the component level and (2) full application is very expensive. Basing logic trees on the probability of failure is a problem because available component probability data are specific to standard conditions and extrapolation techniques cannot be used to modify the data for particular applications.

FAULT-TREE ANALYSIS

Fault-tree analysis is a method of analyzing system reliability and safety. It provides an objective basis for analyzing system design, justifying system changes, performing trade-off studies, analyzing common failure modes, and demonstrating compliance with safety and environment requirements. It is different from a failure mode and effect analysis in that it is restricted to identifying system elements and events that lead to one particular undesired event. Figure 2–2 shows the steps involved in performing a fault-tree analysis.

Many reliability techniques are inductive and concerned primarily with ensuring that hardware accomplishes its intended functions. Fault-tree analysis is a detailed *deductive* analysis that usually requires considerable information about the system. It ensures that all critical aspects of a system are identified and controlled. This method represents graphically the Boolean logic associated with a particular system failure,

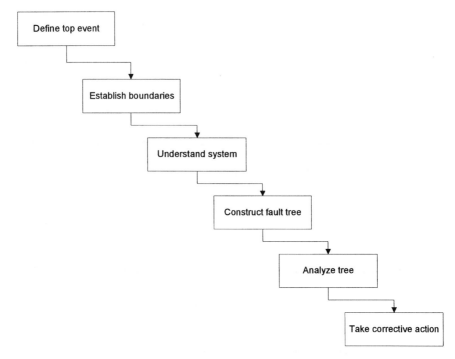

Figure 2–2 Typical fault-tree process.

called the *top event*, and basic failures or causes, called *primary events*. Top events can be broad, all-encompassing system failures or specific component failures.

Fault-tree analysis provides options for performing qualitative and quantitative reliability analysis. It helps the analyst understand system failures deductively and points out the aspects of a system that are important with respect to the failure of interest. The analysis provides insight into system behavior.

A fault-tree model graphically and logically presents the various combinations of possible events occurring in a system that lead to the top event. The term *event* denotes a dynamic change of state that occurs in a system element, which includes hardware, software, human, and environmental factors. A *fault event* is an abnormal system state. A *normal event* is expected to occur.

The structure of a fault tree is shown in Figure 2–3. The undesired event appears as the top event and is linked to more basic fault events by event statements and logic gates.

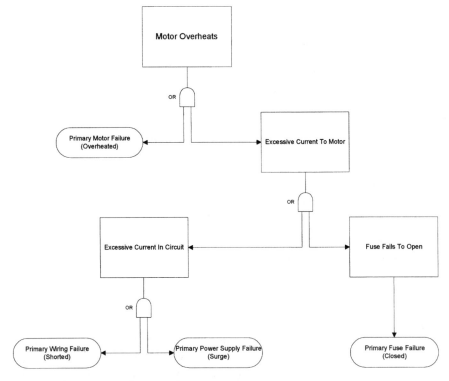

Figure 2–3 Example of a fault-tree logic tree.

CAUSE-AND-EFFECT ANALYSIS

Cause-and-effect analysis is a graphical approach to failure analysis. This also is referred to as *fishbone analysis*, a name derived from the fish-shaped pattern used to plot the relationship between various factors that contribute to a specific event. Typically, fishbone analysis plots four major classifications of potential causes (i.e., human, machine, material, and method) but can include any combination of categories. Figure 2–4 illustrates a simple analysis.

Like most of the failure analysis methods, this approach relies on a logical evaluation of actions or changes that lead to a specific event, such as machine failure. The only difference between this approach and other methods is the use of the fish-shaped graph to plot the cause-effect relationship between specific actions, or changes, and the end result or event.

This approach has one serious limitation. *The fishbone graph provides no clear sequence of events that leads to failure.* Instead, it displays all the possible causes that

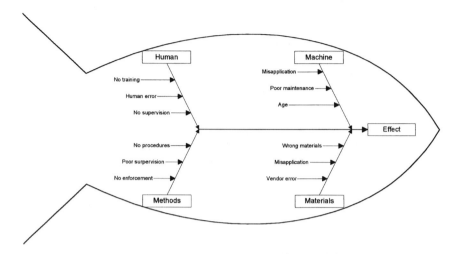

Figure 2–4 Typical fishbone diagram plots four categories of causes.

may have contributed to the event. While this is useful, it does not isolate the specific factors that caused the event. Other approaches provide the means to isolate specific changes, omissions, or actions that caused the failure, release, accident, or other event being investigated.

SEQUENCE-OF-EVENTS ANALYSIS

A number of software programs (e.g., Microsoft's Visio) can be used to generate a *sequence-of-events diagram*. As part of the RCFA program, select appropriate software to use, develop a standard format (see Figure 2–5), and be sure to include each event that is investigated in the diagram.

Using such a diagram from the start of an investigation helps the investigator organize the information collected, identify missing or conflicting information, improve his or her understanding by showing the relationship between events and the incident, and highlight potential causes of the incident.

The sequence-of-events diagram should be a dynamic document generated soon after a problem is reported and continually modified until the event is fully resolved. Figure 2–6 is an example of such a diagram.

Proper use of this graphical tool greatly improves the effectiveness of the problem-solving team and the accuracy of the evaluation. To achieve maximum benefit from

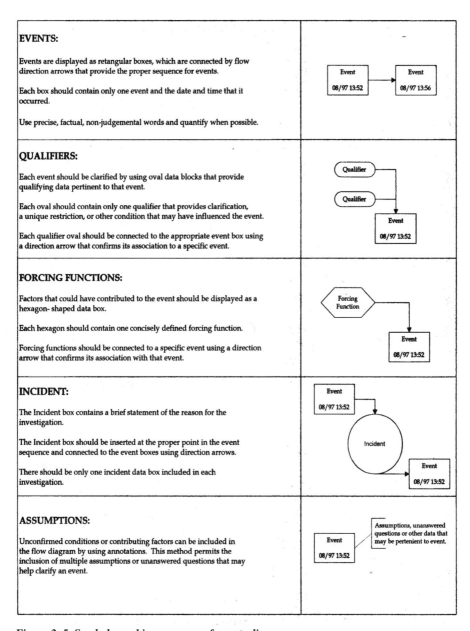

EVENTS:

Events are displayed as retangular boxes, which are connected by flow direction arrows that provide the proper sequence for events.

Each box should contain only one event and the date and time that it occurred.

Use precise, factual, non-judgemental words and quantify when possible.

QUALIFIERS:

Each event should be clarified by using oval data blocks that provide qualifying data pertinent to that event.

Each oval should contain only one qualifier that provides clarification, a unique restriction, or other condition that may have influenced the event.

Each qualifier oval should be connected to the appropriate event box using a direction arrow that confirms its association to a specific event.

FORCING FUNCTIONS:

Factors that could have contributed to the event should be displayed as a hexagon- shaped data box.

Each hexagon should contain one concisely defined forcing function.

Forcing functions should be connected to a specific event using a direction arrow that confirms its association with that event.

INCIDENT:

The Incident box contains a brief statement of the reason for the investigation.

The Incident box should be inserted at the proper point in the event sequence and connected to the event boxes using direction arrows.

There should be only one incident data box included in each investigation.

ASSUMPTIONS:

Unconfirmed conditions or contributing factors can be included in the flow diagram by using annotations. This method permits the inclusion of multiple assumptions or unanswered questions that may help clarify an event.

Figure 2–5 Symbols used in sequence-of-events diagram.

this technique, be consistent and thorough when developing the diagram. The following guidelines should be considered when generating a sequence-of-events diagram: Use a logical order, describe events in active rather than passive terms, be precise, and define or qualify each event or forcing function.

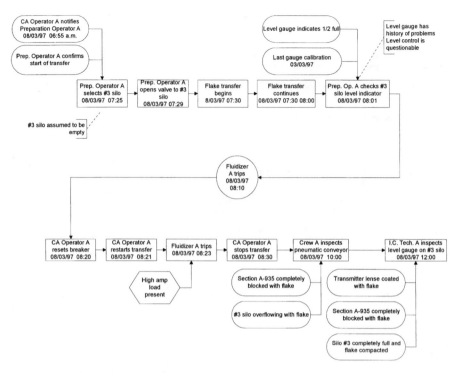

Figure 2–6 Typical sequence-of-events diagram.

In the example illustrated in Figure 2–6, repeated trips of the fluidizer used to transfer flake from the Cellulose Acetate (CA) Department to the preparation area triggered an investigation. The diagram shows each event that led to the initial and second fluidizer trip. The final event, the silo inspection, indicated that the root cause of the problem was failure of the level-monitoring system. Because of this failure, Operator A over-filled the silo. When this happened, the flake compacted in the silo and backed up in the pneumatic-conveyor system. This backup plugged an entire section of the pneu-matic-conveyor piping, which resulted in an extended production outage while the plug was removed.

Logical Order

Show events in a logical order from the beginning to the end of the sequence. Initially, the sequence-of-events diagram should include all pertinent events, including those that cannot be confirmed. As the investigation progresses, it should be refined to show only those events that are confirmed to be relevant to the incident.

Active Descriptions

Event boxes in a sequence-of-events diagram should contain action steps rather than passive descriptions of the problem. For example, the event should read: "Operator A pushes pump start button" not "The wrong pump was started." As a general rule, only one subject and one verb should be used in each event box. Rather than "Operator A pushed the pump stop button and verified the valve line-up," two event boxes should be used. The first box should say "Operator A pushed the pump stop button" and the second should say "Operator A verified valve line-up."

Do not use people's names on the diagram. Instead use job functions or assign a code designator for each person involved in the event or incident. For example, three operators should be designated Operator A, Operator B, and Operator C.

Be Precise

Precisely and concisely describe each event, forcing function, and qualifier. If a concise description is not possible and assumptions must be provided for clarity, include them as annotations. This is described in Figure 2–5 and illustrated in Figure 2–6. As the investigation progresses, each assumption and unconfirmed contributor to the event must be either confirmed or discounted. As a result, each event, function, or qualifier generally will be reduced to a more concise description.

Define Events and Forcing Functions

Qualifiers that provide all confirmed background or support data needed to accurately define the event or forcing function should be included in a sequence-of-events diagram. For example, each event should include date and time qualifiers that fix the time frame of the event.

When confirmed qualifiers are unavailable, assumptions may be used to define unconfirmed or perceived factors that may have contributed to the event or function. However, every effort should be made during the investigation to eliminate the assumptions associated with the sequence-of-events diagram and replace them with known facts.

3

ROOT CAUSE FAILURE
ANALYSIS METHODOLOGY

RCFA is a logical sequence of steps that leads the investigator through the process of isolating the facts surrounding an event or failure. Once the problem has been fully defined, the analysis systematically determines the best course of action that will resolve the event and assure that it is not repeated. Because of the cost associated with performing such an analysis, care should be exercised before an investigation is undertaken.

The first step in this process is obtaining a clear definition of the potential problem or event. The logic tree illustrated in Figure 3–1 should be followed for the initial phase of the evaluation.

REPORTING AN INCIDENT OR PROBLEM

The investigator seldom is present when an incident or problem occurs. Therefore, the first step is the initial notification that an incident or problem has taken place. Typically, this report will be verbal, a brief written note, or a notation in the production log book. In most cases, the communication will not contain a complete description of the problem. Rather, it will be a very brief description of the perceived symptoms observed by the person reporting the problem.

Symptoms and Boundaries

The most effective means of problem or event definition is to determine its *real* symptoms and establish limits that bound the event. At this stage of the investigation, the task can be accomplished by an interview with the person who first observed the problem.

Perceived Causes of Problem

At this point, each person interviewed will have a definite opinion about the incident, and will have his or her description of the event and an absolute reason for the occurrence. In

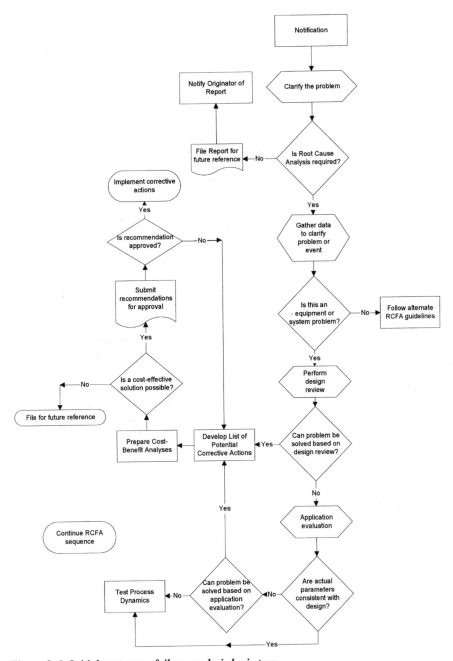

Figure 3–1 Initial root cause failure analysis logic tree.

many cases, these perceptions are totally wrong, but they cannot be discounted. Even though many of the opinions expressed by the people involved with or reporting an event may be invalid, do not discount them without investigation. Each opinion should be recorded and used as part of the investigation. In many cases, one or more of the opinions will hold the key to resolution of the event. The following are some examples where the initial perception was incorrect.

One example of this phenomenon is a reported dust collector baghouse problem. The initial report stated that dust-laden air was being vented from the baghouses on a random, yet recurring, basis. The person reporting the problem was convinced that chronic failure of the solenoid-actuated pilot valves controlling the blow-down of the baghouse, without a doubt, was the cause. However, a quick design review found that the solenoid-controlled valves *normally are closed*. This type of solenoid valve *cannot fail* in the *open* position and, therefore, could not be the source of the reported events.

A conversation with a process engineer identified the diaphragms used to seal the blow-down tubes as a potential problem source. This observation, coupled with inadequate plant air, turned out to be the root cause of the reported problem.

Another example illustrating preconceived opinions is the catastrophic failure of a Hefler chain conveyor. In this example, all the bars on the left side of the chain were severely bent before the system could be shut down. Even though no foreign object such as a bolt was found, this was assumed to be the cause for failure. From the evidence, it was clear that some obstruction had caused the conveyor damage, but the more important question was, Why did it happen?

Hefler conveyors are designed with an intentional failure point that should have prevented the extensive damage caused by this event. The main drive-sprocket design includes a *shear pin* that generally prevents this type of catastrophic damage. Why did the conveyor fail? Because the shear pins had been removed and replaced with Grade-5 bolts.

Event-Reporting Format

One factor that severely limits the effectiveness of RCFA is the absence of a formal event-reporting format. The use of a format that completely bounds the potential problem or event greatly reduces the level of effort required to complete an analysis. A form similar to the one shown in Figure 3–2 provides the minimum level of data needed to determine the effort required for problem resolution.

INCIDENT CLASSIFICATION

Once the incident has been reported, the next step is to identify and classify the type of problem. Common problem classifications are equipment damage or failure, operating performance, economic performance, safety, and regulatory compliance.

INCIDENT REPORTING FORM

Date:

Reported By:

Description of Incident:

Specific Location and Equipment/System Effected:

When Did Incident Occur:

Who Was Involved:

What Is Probable Cause:

What Corrective Actions Taken:

Was Personal Injury Involved: ☐ Yes ☐ No

Was Reportable Release Involved: ☐ Yes ☐ No

Incident Classification: ☐ Equipment Failure ☐ Regulatory Compliance
☐ Accident/Injury ☐ Performance Deviation

Signature

Figure 3–2 Typical incident-reporting form.

Classifying the event as a particular problem type allows the analyst to determine the best method to resolve the problem. Each major classification requires a slightly different RCFA approach, as shown in Figure 3–3.

Note, however, that initial classification of the event or problem typically is the most difficult part of a RCFA. Too many plants lack a formal tracking and reporting system that accurately detects and defines deviations from optimum operation condition.

Equipment Damage or Failure

A major classification of problems that often warrants RCFA are those events associated with the failure of critical production equipment, machinery, or systems. Typically, any incident that results in partial or complete failure of a machine or process system warrants a RCFA. This type of incident can have a severe, negative impact on plant performance. Therefore, it often justifies the effort required to fully evaluate the event and to determine its root cause.

Events that result in physical damage to plant equipment or systems are the easiest to classify. Visual inspection of the failed machine or system component usually provides clear evidence of its failure mode. While this inspection usually will not resolve the reason for failure, the visible symptoms or results will be evident. The events that also meet other criteria (e.g., safety, regulatory, or financial impact) should be investigated automatically to determine the actual or potential impact on plant performance, including equipment reliability.

Figure 3–3 Initial event classification.

In most cases, the failed machine must be replaced immediately to minimize its impact on production. If this is the case, evaluating the system surrounding the incident may be beneficial.

Operating Performance

Deviations in operating performance may occur without the physical failure of critical production equipment or systems. Chronic deviations may justify the use of RCFA as a means of resolving the recurring problem.

Generally, chronic product quality and capacity problems require a full RCFA. However, care must be exercised to ensure that these problems are recurring and have a significant impact on plant performance before using this problem-solving technique.

Product Quality

Deviations in first-time-through product quality are prime candidates for RCFA, which can be used to resolve most quality-related problems. However, the analysis should not be used for all quality problems. Nonrecurring deviations or those that have no significant impact on capacity or costs are not cost-effective applications.

Capacity Restrictions

Many of the problems or events that occur affect a plant's ability to consistently meet expected production or capacity rates. These problems may be suitable for RCFA, but further evaluation is recommended before beginning an analysis. After the initial investigation, if the event can be fully qualified and a cost-effective solution not found, then a full analysis should be considered. Note that an analysis normally is not performed on random, nonrecurring events or equipment failures.

Economic Performance

Deviations in economic performance, such as high production or maintenance costs, often warrant the use of RCFA. The decision tree and specific steps required to resolve these problems vary depending on the type of problem and its forcing functions or causes.

Safety

Any event that has a potential for causing personal injury should be investigated immediately. While events in this classification may not warrant a full RCFA, they must be resolved as quickly as possible.

Isolating the root cause of injury-causing accidents or events generally is more difficult than for equipment failures and requires a different problem-solving approach. The primary reason for this increased difficulty is that the cause often is subjective.

Regulatory Compliance

Any regulatory compliance event can have a potential impact on the safety of workers, the environment, as well as the continued operation of the plant. Therefore, any event that results in a violation of environmental permits or other regulatory-compliance guidelines (e.g., Occupational Safety and Health Administration, Environmental Protection Agency, and state regulations) should be investigated and resolved as quickly as possible. Since all releases and violations must be reported—and they have a potential for curtailed production or fines or both—this type of problem must receive a high priority.

DATA GATHERING

The data-gathering step should clarify the reported event or problem. This phase of the evaluation includes interviews with appropriate personnel, collecting physical evidence, and conducting other research, such as performing a sequence-of-events analysis, which is needed to provide a clear understanding of the problem. Note that this section focuses primarily on equipment damage or failure incidents.

Interviews

The interview process is the primary method used to establish actual boundary conditions of an incident and is a key part of any investigation. It is crucial for the investigator to be a good listener with good diplomatic and interviewing skills.

For significant incidents, all key personnel must be interviewed to get a complete picture of the event. In addition to those directly involved in the event or incident, individuals having direct or indirect knowledge that could help clarify the event should be interviewed. The following is a partial list of interviewees:

- All personnel directly involved with the incident (be sure to review any written witness statements).
- Supervisors and managers of those involved in the incident (including contractor management).
- Personnel not directly involved in the incident but who have similar background and experience.
- Applicable technical experts, training personnel, and equipment vendors, suppliers, or manufacturers.

Note that it is extremely important for the investigator to convey the message that the purpose of an interview is fact finding *not* fault finding. The investigator's job is simply to find out what actually happened and why it happened. *It is important for the interviewer to clearly define the reason for the evaluation to the interviewee at the beginning of the interview process.* Plant personnel must understand and believe that the reason for the evaluation is to find the problem. If they believe that the process is intended to fix blame, little benefit can be derived.

It also is necessary to verify the information derived from the interview process. One means of verification is visual observation of the actual practices used by the production and maintenance teams assigned to the area being investigated.

Questions to Ask

To listen more effectively the interviewer must be prepared for the interview, and preparation helps avoid wasting time. Prepared questions or a list of topics to discuss helps keep the interview on track and prevents the interviewer from forgetting to ask questions on key topics. Figure 3–4 is a flow sheet summarizing the interview process. Each interview should be conducted to obtain clear answers to the following questions:

- What happened?
- Where did it happen?

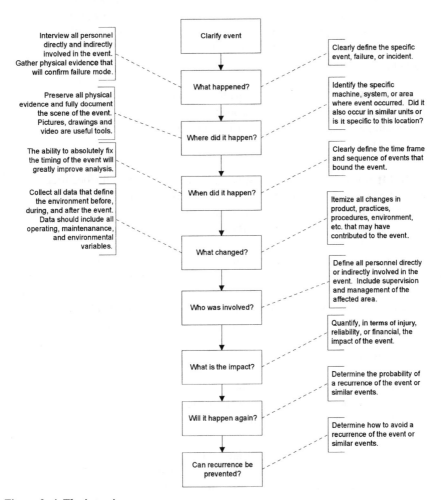

Figure 3–4 The interview process.

- When did it happen?
- What changed?
- Who was involved?
- Why did it happen?
- What is the impact?
- Will it happen again?
- How can recurrence be prevented?

What Happened?   Clarifying what actually happened is an essential requirement of RCFA. As discussed earlier, the natural tendency is to give perceptions rather than to carefully define the actual event. It is important to include as much detail as the facts and available data permit.

Where Did It Happen?   A clear description of the exact location of the event helps isolate and resolve the problem. In addition to the location, determine if the event also occurred in similar locations or systems. If similar machines or applications are eliminated, the event sometimes can be isolated to one, or a series of, forcing function(s) totally unique to the location.

For example, if Pump A failed and Pumps B, C, and D in the same system did not, this indicates that the reason for failure is probably unique to Pump A. If Pumps B, C, and D exhibit similar symptoms, however, it is highly probable that the cause is systemic and common to all the pumps.

When Did It Happen?   Isolating the specific time that an event occurred greatly improves the investigator's ability to determine its source. When the actual time frame of an event is known, it is much easier to quantify the process, operations, and other variables that may have contributed to the event.

However, in some cases (e.g., product-quality deviations), it is difficult to accurately fix the beginning and duration of the event. Most plant-monitoring and tracking records do not provide the level of detail required to properly fix the time of this type of incident. In these cases, the investigator should evaluate the operating history of the affected process area to determine if a pattern can be found that properly fixes the event's time frame. This type of investigation, in most cases, will isolate the timing to events such as the following:

- Production of a specific product.
- Work schedule of a specific operating team.
- Changes in ambient environment.

What Changed?   Equipment failures and major deviations from acceptable performance levels do not just happen. In every case, specific variables, singly or in combination, caused the event to occur. Therefore, it is essential that any changes that occurred in conjunction with the event be defined.

No matter what the event is (i.e., equipment failure, environmental release, accident, etc.), the evaluation must quantify all the variables associated with the event. These data should include the operating setup; product variables, such as viscosity, density, flow rates, and so forth; and the ambient environment. If available, the data also should include any predictive-maintenance data associated with the event.

Who Was Involved? The investigation should identify all personnel involved, directly or indirectly, in the event. Failures and events often result from human error or inadequate skills. However, remember that the purpose of the investigation is to resolve the problem, not to place blame.

All comments or statements derived during this part of the investigation should be impersonal and totally objective. All references to personnel directly involved in the incident should be assigned a *code number or other identifier*, such as Operator A or Maintenance Craftsman B. This approach helps reduce fear of punishment for those directly involved in the incident. In addition, it reduces prejudice or preconceived opinions about individuals within the organization.

Why Did It Happen? If the preceding questions are fully answered, it may be possible to resolve the incident with no further investigation. However, exercise caution to ensure that the real problem has been identified. It is too easy to address the symptoms or perceptions without a full analysis.

At this point, generate a list of what may have contributed to the reported problem. The list should include *all* factors, both real and assumed. This step is critical to the process. In many cases, a number of factors, many of them trivial, combine to cause a serious problem.

All assumptions included in this list of possible causes should be clearly noted, as should the causes that are proven. A *sequence-of-events analysis* provides a means for separating fact from fiction during the analysis process.

What Is the Impact? The evaluation should quantify the impact of the event before embarking on a full RCFA. Again, not all events, even some that are repetitive, warrant a full analysis. This part of the investigation process should be as factual as possible. Even though all the details are unavailable at this point, attempt to assess the real or potential impact of the event.

Will It Happen Again? If the preliminary interview determines that the event is nonrecurring, the process may be discontinued at this point. However, a thorough review of the historical records associated with the machine or system involved in the incident should be conducted before making this decision. Make sure that it truly is a nonrecurring event before discontinuing the evaluation.

All reported events should be recorded and the files maintained for future reference. For incidents found to be nonrecurring, a file should be established that retains all the data and information developed in the preceding steps. Should the event or a similar one occur again, these records are an invaluable investigative tool.

A full investigation should be conducted on any event that has a history of periodic recurrence, or a high probability of recurrence, and a significant impact in terms of injury, reliability, or economics. In particular, all incidents that have the potential for personal injury or regulatory violation should be investigated.

How Can Recurrence Be Prevented? Although this is the next logical question to ask, it generally cannot be answered until the entire RCFA is completed. Note, however, that if this analysis determines it is not economically feasible to correct the problem, plant personnel may simply have to learn to minimize the impact.

Types of Interviews

One of the questions to answer in preparing for an interview is "What type of interview is needed for this investigation?" Interviews can be grouped into three basic types: one-on-one, two-on-one, and group meetings.

One-on-One The simplest interview to conduct is that where the investigator interviews each person necessary to clarify the event. This type of interview should be held in a private location with no distractions. In instances where a field walk-down is required, the interview may be held in the employee's work space.

Two-on-One When controversial or complex incidents are being investigated, it may be advisable to have two interviewers present when meeting with an individual. With two investigators, one can ask questions while the other records information. The interviewers should coordinate their questioning and avoid overwhelming or intimidating the interviewee.

At the end of the interview, the interviewers should compare their impressions of the interview and reach a consensus on their views. The advantage of the two-on-one interview is that it should eliminate any personal perceptions of a single interviewer from the investigation process.

Group Meeting A group interview is advantageous in some instances. This type of meeting, or group problem-solving exercise, is useful for obtaining an interchange of ideas from several disciplines (i.e., maintenance, production, engineering, etc.). Such an interchange may help resolve an event or problem.

This approach also can be used when the investigator has completed his or her evaluation and wants to review the findings with those involved in the incident. The investigator might consider interviews with key witnesses before the group meeting to verify the sequence of events and the conclusions before presenting them to the larger group. The investigator must act as facilitator in this problem-solving process and use a sequence-of-events diagram as the working tool for the meeting.

Group interviews cannot be used in a hostile environment. If the problem or event is controversial or political, this type of interview process is not beneficial. The personal agendas of the participants generally preclude positive results.

Collecting Physical Evidence

The first priority when investigating an event involving equipment damage or failure is to preserve physical evidence. Figure 3–5 is a flow diagram illustrating the steps involved in an equipment-failure investigation. This effort should include all tasks and activities required to fully evaluate the failure mode and determine the specific boundary conditions present when the failure occurred.

If possible, the failed machine and its installed system should be isolated from service until a full investigation can be conducted. On removal from service, the failed machine and all its components should be stored in a secure area until they can be fully inspected and appropriate tests conducted.

If this approach is not practical, the scene of the failure should be fully documented before the machine is removed from its installation. Photographs, sketches, and the instrumentation and control settings should be fully documented to ensure that all data are preserved for the investigating team. All automatic reports, such as those generated by the Level I computer-monitoring system, should be obtained and preserved.

The legwork required to collect information and physical evidence for the investigation can be quite extensive. The following is a partial list of the information that should be gathered:

- Currently approved standard operating (SOP) and maintenance (SMP) procedures for the machine or area where the event occurred.
- Company policies that govern activities performed during the event.
- Operating and process data (e.g., strip charts, computer output, and data-recorder information).
- Appropriate maintenance records for the machinery or area involved in the event.
- Copies of log books, work packages, work orders, work permits, and maintenance records; equipment-test results, quality-control reports; oil and lubrication analysis results; vibration signatures; and other records.
- Diagrams, schematics, drawings, vendor manuals, and technical specifications, including pertinent design data for the system or area involved in the incident.
- Training records, copies of training courses, and other information that shows skill levels of personnel involved in the event.
- Photographs, videotape, or diagrams of the incident scene.
- Broken hardware (e.g., ruptured gaskets, burned leads, blown fuses, failed bearings).

- Environmental conditions when the event occurred. These data should be as complete and accurate as possible.
- Copies of incident reports for similar prior events and history or trend information for the area involved in the current incident.

Figure 3–5 Flow diagram for equipment failure investigation.

Analyze Sequence of Events

Performing a sequence-of-events analysis and graphically plotting the actions leading up to and following an event, accident, or failure helps visualize what happened. It is important to use such a diagram from the start of an investigation. This not only helps with organizing the information but also in identifying missing or conflicting data, showing the relationship between events and the incident, and highlighting potential causes of the incident.

DESIGN REVIEW

It is essential to clearly understand the design parameters and specifications of the systems associated with an event or equipment failure. Unless the investigator understands precisely what the machine or production system was designed to do and its inherent limitations, it is impossible to isolate the root cause of a problem or event. The data obtained from a design review provide a baseline or reference, which is needed to fully investigate and resolve plant problems.

The objective of the design review is to establish the specific operating characteristics of the machine or production system involved in the incident. The evaluation should clearly define the specific function or functions that each machine and system was designed to perform. In addition, the review should establish the acceptable operating envelope, or range, that the machine or system can tolerate without a measurable deviation from design performance.

The logic used for a comprehensive review is similar to that of a failure modes and effects analysis and a fault-tree analysis in that it is intended to identify the contributing variables. Unlike these other techniques, which use complex probability tables and break down each machine to the component level, RCFA takes a more practical approach. The technique is based on readily available, application-specific data to determine the variables that may cause or contribute to an incident.

While the level of detail required for a design review varies depending on the type of event, this step cannot be omitted from any investigation. In some instances, the process may be limited to a cursory review of the vendor's *operating and maintenance (O&M) manual* and performance specifications. In others, a full evaluation that includes all procurement, design, and operations data may be required.

Minimum Design Data

In many cases, the information required can be obtained from four sources: equipment nameplates, procurement specifications, vendor specifications, and the O&M manuals provided by the vendors.

If the investigator has a reasonable understanding of machine dynamics, a thorough design review for relatively simple production systems (e.g., pump transfer system) can be accomplished with just the data provided in these four documents. If the investigator lacks a basic knowledge of machine dynamics, review Part Two of this book, Equipment Design Evaluation Guides.

Special attention should be given to the vendor's troubleshooting guidelines. These suggestions will provide insight into the more common causes for abnormal behavior and failure modes.

Equipment Nameplate Data

Most of the machinery, equipment, and systems used in process plants have a permanently affixed nameplate that defines their operating envelope. For example, a centrifugal pump's nameplate typically includes its flow rate, total discharge pressure, specific gravity, impeller diameter, and other data that define its design operating characteristics. These data can be used to determine if the equipment is suitable for the application and if it is operating within its design envelope.

Procurement Specifications

Procurement specifications normally are prepared for all capital equipment as part of the purchasing process. These documents define the specific characteristics and operating envelope requested by the plant engineering group. The specifications provide information useful for evaluating the equipment or system during an investigation.

When procurement specifications are unavailable, purchasing records should describe the equipment and provide the system envelope. Although such data may be limited to a specific type or model of machine, it generally is useful information.

Vendor Specifications

For most equipment procured as a part of capital projects, a detailed set of vendor specifications should be available. Generally, these specifications were included in the vendor's proposal and confirmed as part of the deliverables for the project. Normally, these records are on file in two different departments: purchasing and plant engineering.

As part of the design review, the vendor and procurement specifications should be carefully compared. *Many of the chronic problems that plague plants are a direct result of vendor deviations from procurement specifications.* Carefully comparing these two documents may uncover the root cause of chronic problems.

Operating and Maintenance Manuals

O&M manuals are one of the best sources of information. In most cases, these documents provide specific recommendations for proper operation and maintenance of the machine, equipment, or system. In addition, most of these manuals provide specific troubleshooting guides that point out many of the common problems that may occur. A thorough review of these documents is essential before beginning the RCFA. The information provided in these manuals is essential to effective resolution of plant problems.

Objectives of the Review

The objective of the design review is to determine the design limitations, acceptable operating envelope, probable failure modes, and specific indices that quantify the actual operating condition of the machine, equipment, or process system being investigated. At a minimum, the evaluation should determine the design function and specifically what the machine or system was designed to do. The review should clearly define the specific functions of the system and its components.

To fully define machinery, equipment, or system functions, a description should include incoming and output product specifications, work to be performed, and acceptable operating envelopes. For example, a centrifugal pump may be designed to deliver 1,000 gallons per minute of water having a temperature of 100°F and a discharge pressure of 100 pounds per square inch.

Incoming-Product Specifications

Machine and system functions depend on the incoming product to be handled. Therefore, the design review must establish the incoming product boundary conditions used in the design process. In most cases, these boundaries include temperature range, density or specific gravity, volume, pressure, and other measurable parameters. These boundaries determine the amount of work the machine or system must provide.

In some cases, the boundary conditions are absolute. In others, there is an acceptable range for each of the variables. The review should clearly define the allowable boundaries used for the system's design.

Output-Product Specifications

Assuming the incoming product boundary conditions are met, the investigation should determine what output the system was designed to deliver. As with the incoming product, the output from the machine or system can be bound by specific, measurable parameters. Flow, pressure, density, and temperature are the common measures of output product. However, depending on the process, there may be others.

Work to Be Performed

This part of the design review should determine the measurable work to be performed by the machine or system. Efficiency, power usage, product loss, and similar parameters are used to define this part of the review. The actual parameters will vary depending on the machine or system. In most cases, the original design specifications will provide the proper parameters for the system under investigation.

Acceptable Operating Envelope

The final part of the design review is to define the acceptable operating envelope of the machine or system. Each machine or system is designed to operate within a specific range, or operating envelope. This envelope includes the maximum variation in incoming product, startup ramp rates and shut-down speeds, ambient environment, and a variety of other parameters.

APPLICATION/MAINTENANCE REVIEW

The next step in the RCFA is to review the application to ensure that the machine or system is being used in the proper application. The data gathered during the design review should be used to verify the application. The maintenance record also should be reviewed.

In plants where multiple products are produced by the machine or process system being investigated, it is essential that the full application range be evaluated. The evaluation must include all variations in the operating envelope over the full range of products being produced. The reason this is so important is that many of the problems that will be investigated are directly related to one or more process setups that may be unique to that product. Unless the full range of operation is evaluated, there is a potential that the root cause of the problem will be missed.

Factors to evaluate in an application/maintenance review include installation, operating envelope, operating procedures and practices (i.e., standard procedures versus actual practices), maintenance history, and maintenance procedures and practices.

Installation

Each machine and system has specific installation criteria that must be met before acceptable levels of reliability can be achieved and sustained. These criteria vary with the type of machine or system and should be verified as part of the RCFA.

Using the information developed as part of the design review, the investigator or other qualified individuals should evaluate the actual installation of the machine or system being investigated. At a minimum, a thorough visual inspection of the machine and its related system should be conducted to determine if improper installation is contributing to the problem. The installation requirements will vary depending on the type of machine or system.

Photographs, sketches, or drawings of the actual installation should be prepared as part of the evaluation. They should point out any deviations from acceptable or recommended installation practices as defined in the reference documents and good engineering practices. This data can be used later in the RCFA when potential corrective actions are considered.

Operating Envelope

Evaluating the actual operating envelope of the production system associated with the investigated event is more difficult. The best approach is to determine all variables and limits used in normal production. For example, define the full range of operating speeds, flow rates, incoming product variations, and the like normally associated with the system. In variable-speed applications, determine the minimum and maximum ramp rates used by the operators.

Operating Procedures and Practices

This part of the application/maintenance review consists of evaluating the standard operating procedures as well as the actual operating practices. Most production areas maintain some historical data that track its performance and practices. These records may consist of log books, reports, or computer data. These data should be reviewed to determine the actual production practices that are used to operate the machine or system being investigated.

Systems that use a computer-based monitoring and control system will have the best database for this part of the evaluation. Many of these systems automatically store and, in some cases, print regular reports that define the actual process setups for each type of product produced by the system. This invaluable source of information should be carefully evaluated.

Standard Operating Procedures

Evaluate the standard operating procedures for the affected area or system to determine if they are consistent and adequate for the application. Two reference sources, the design review report and vendor's O&M manuals, are required to complete this task.

In addition, evaluate SOPs to determine if they are usable by the operators. Review organization, content, and syntax to determine if the procedure is correct and understandable.

Setup Procedures

Special attention should be given to the setup procedures for each product produced by a machine or process system. Improper or inconsistent system setup is a leading cause of poor product quality, capacity restrictions, and equipment unreliability. The procedures should provide clear, easy to understand instructions that ensure accurate, repeatable setup for each product type. If they do not, the deviations should be noted for further evaluation.

Transient Procedures

Transient procedures, such as startup, speed change, and shutdown, also should be carefully evaluated. These are the predominant transients that cause deviations in quality and capacity and have a direct impact on equipment reliability. The procedures should be evaluated to ensure that they do not violate the operating envelope or vendor's recommendations. All deviations must be clearly defined for further evaluation.

Operating Practices

This part of the evaluation should determine if the SOPs were understood and followed before and during the incident or event. *The normal tendency of operators is to shortcut procedures, which is a common reason for many problems.* In addition, unclear procedures lead to misunderstanding and misuse.

Therefore, the investigation must fully evaluate the actual practices that the production team uses to operate the machine or system. The best way to determine compliance with SOPs is to have the operator(s) list the steps used to run the system or machine being investigated. This task should be performed without referring to the SOP manual. The investigator should lead the operator(s) through the process and use their input to develop a sequence diagram.

After the diagram is complete, compare it to the SOPs. If the operator's actual practices are not the same as those described in the SOPs, the procedures may need to be upgraded or the operators may need to be retrained.

Maintenance History

A thorough review of the maintenance history associated with the machine or system is essential to the RCFA process. One question that must be answered is, "Will this happen again?" A review of the maintenance history may help answer this question. The level of accurate maintenance data available will vary greatly from plant to plant. This may hamper the evaluation, but it is necessary to develop as clear a picture as possible of the system's maintenance history.

A complete history of the scheduled and actual maintenance, including inspections and lubrication, should be developed for the affected machine, system, or area. The primary details that are needed include frequency and types of repair, frequency and types of preventive maintenance, failure history, and any other facts that will help in the investigation.

Maintenance Procedures and Practices

A complete evaluation of the standard maintenance procedures and actual practices should be conducted. The procedures should be compared with maintenance requirements defined by both the design review and the vendor's O&M manuals. Actual maintenance practices can be determined in the same manner as described earlier or by visual observation of similar repairs. This task should determine if the SMPs are followed consistently by all maintenance personnel assigned to or involved with the area being investigated. Special attention should be given to the routine tasks, such as lubrication, adjustments, and other preventive tasks. Determine if these procedures are being performed in a timely manner and if proper techniques are being used.

OBSERVATIONS AND MEASUREMENTS

If the application and maintenance reviews are not conclusive, additional evaluation and testing may be necessary. The techniques discussed in this section provide ways to quantify deviations from acceptable operating and maintenance practices. By using these techniques, the investigating team can accurately establish deviations that may be contributing to the problem.

After establishing potential deviations in the operating envelope or other variations that may have contributed to the event, the next step is to verify that these factors were actual contributors. The best method of confirmation is to physically measure the operating dynamics of the machine or production system associated with the event.

Evaluating any event or incident that results in partial or complete failure of a machine, piece of equipment, or system should include a thorough analysis of the failed parts or components. The actual methods used will vary depending on the failure mode and the type of machine or component that failed.

In most cases, failed components must be replaced immediately to minimize the impact on production. If this is the case, an evaluation of the system that surrounded the incident may be beneficial. If this testing can be conducted shortly after replacing the failed machine, there may be residual evidence that will aid in the investigation. This is especially true in those cases where variations in operating parameters, such as product variations and operating practices, contributed to the failure.

Many types of measurements can be taken and observations made, but the most effective include vibration analysis, quantification of process parameters, and visual inspection. Vibration analysis, coupled with direct recording of process variables, provides the most effective means of determining their effect. A full multichannel analysis of the replacement machine and its installed system provides the data required to evaluate the potential contribution of the system to the failure.

Vibration Analysis

Vibration analysis is perhaps the best method of measuring the actual operating dynamics of mechanical systems. This technique provides the means of quantifying a variety of factors that may contribute to the event—even when equipment failure has not occurred.

If variations in incoming product, operating or maintenance practices, ambient environment, or a myriad or other variables are believed to be a contributing factor, vibration tests can measure the real effect of each variable.

It is necessary to develop a test sequence that isolates and directly compares each suspect variable. For example, a series of vibration tests can be conducted to measure the actual change in the machine or process system's response to discrete changes in incoming-product parameters, such as viscosity or temperature. Another sequence of tests can be conducted to measure the effects of variations in operating practices or ambient environment.

All vibration testing should be based on a well-thought-out test plan that defines its purpose and the methods to be used. Good, detailed documentation is essential. Each test in the sequence must be fully detailed to clearly define the exact parameters to use during data acquisition. This documentation is essential for proper analysis and comparison of the data.

In most cases, process or ambient environment data must be acquired along with the vibration data. Since these tests are designed to quantify the effects of changes in the variable, it is necessary to verify the actual change caused by each.

Real-time data can be used to accurately quantify the cause-and-effect relationship between suspected variables and the operating dynamics of the machine or system. Multichannel data-acquisition techniques provide the best means of acquiring this type of data. Multichannel techniques record all variables at the same time, eliminating any transients that may be missed with single-channel data acquisition methods.

Process Parameters

Evaluating variations in process parameters, such as pressures flow rate, and temperature, is an effective means of confirming their impact on the production system. In some cases, the Level I computer monitoring systems used to operate the process system provide much of the data needed. The monitoring systems normally acquire, report, and store key process variables throughout the normal production process. This information is useful for both the initial analysis and confirmation testing.

Tests should be developed that measure the impact of process-parameter changes on the system to determine if specific deviations contributed to the event being investigated. In most cases, these tests should be used in conjunction with vibration analysis to measure the cause-and-effect relationship between specific process setups and the operating dynamics of the system.

Visual Inspection

Visual inspection is one tool that *always* should be used. Such inspections should be performed on measurement devices, such as pressure gauges, and failed machine components.

Measurement Devices

Most machines and systems include measurement devices that provide a clear indication of the operating condition. A visual inspection of these devices confirms many of the failure modes that cause process deviations and catastrophic failures.

For example, pressure gauges are a primary tool used to evaluate operating dynamics. A steady pressure reading is a good indication that the equipment (e.g., pump, fan, compressor) is operating normally and within its designed operating envelope. A fluctuating gauge is a good indication that something is wrong. Turbulent flow, cycling or variation in load or demand, and a variety of other process deviations result in a radical fluctuation of the pressure-gauge reading.

Failed Machine Components

Visual inspection of failed machine components can be an effective diagnostic tool. For example, visual inspection of a failed bearing can determine the load zone, wear pattern, and in many cases the probable cause of the failure. The same is true of gears, V-belts, and a variety of other common machine components.

In many cases, visual inspection provides positive clues that point the investigator toward the probable failure mode or root cause. For example, the wear pattern of a rolling-element bearing or gear set provides visual indications of abnormal loading or behavior.

A number of reference books discuss and illustrate the common failure modes of machine components that can be detected using visual observation as the primary diagnostics tool. Most of these books use photographs to clearly display the visual image of various failure modes.

Figure 3–6 illustrates a damaged rolling-element bearing. From the wear pattern, it is clear that the load zone is abnormal. Rather than being centered in the races, the wear pattern is offset. This visual inspection confirms that some forcing function within the machine or its system has caused an abnormal loading of the bearing. Failure modes (e.g., misalignment, aerodynamic or hydraulic instability, and changes in load or speed) are primary sources of this type of abnormal wear pattern.

Figure 3–7 illustrates an abnormal wear pattern on a spur gear caused by too little backlash. The vertical score marks are caused by sliding wear resulting from too little clearance between the gear set.

Wear Particles

Wear-particle analysis is a diagnostic technique that uses visual analysis aided by an electron microscope to evaluate the wear patterns of the metallic solids contained in a

Figure 3–6 Visual profile of abnormal loading of a rolling-element bearing.

machine's lubrication system. Using this technique, a trained analyst can determine the type of wear and its most probable cause. In most cases, this analysis requires the use of an outside laboratory, with costs ranging from about $25 to $50 per sample. However, the knowledge gained from this test is extremely valuable to the RCFA process.

Figure 3–7 Visual profile of badly worn spur gear.

Other

Other testing techniques (e.g., meggering, ultrasonics, acoustic emissions) can be used to measure the cause-and-effect relationship between variables. Each technique is appropriate for specific applications and provides a means of quantifying or proving that specific forcing functions contributed to the event or incident being investigated.

There also are a variety of tests that can be conducted to evaluate the metallurgical properties of the failed component. These include tensile strength, residual stress, and other properties of the component as well as the actual failure mechanism (i.e., stress rupture, compression fracture, etc.).

Unless the plant has a fully functional metallurgical laboratory, these tests may require the use of an outside lab. The cost of the evaluations will vary, depending on the type of component and the extent of the analysis. Used selectively, metallurgical analysis is an effective tool and can be a useful part of a RCFA.

DETERMINING THE ROOT CAUSE

At this point in the investigation, the preceding steps, sequence-of-events diagrams, and logic trees should have identified the forcing functions that may have contributed to the problem. These data should have provided enough insight into the failure for the investigator to develop a list of potential or probable reasons for the failure.

The guidelines for specific machine types, which are provided in Part Two, and the troubleshooting guides in Part Three should be used as part of this activity. The most common failure modes of machinery are defined in the tables in Part Two, which provide the obvious symptoms associated with a machine problem or failure and their common causes. Coupled with the data developed during the investigation, these tables should help in reducing the number of potential causes to no more than *two* or *three*.

A method that can be used to evaluate potential factors that may have caused or contributed to the event is the cause-and-effect analysis. Graphical representation of the variables or factors identified by the investigation using a fishbone diagram provides a means of evaluating potential root cause(s).

The advantage of the fishbone diagram is that it forces the investigator to logically group each of the factors identified during the investigation. This process may automatically eliminate some factors and uncover other issues that must be addressed. Once all the identified factors have been graphically displayed, the investigator or investigating team can systematically evaluate each one.

Analyzing the short list of potential root causes to verify each of the suspect causes is essential. In almost all cases, a relatively simple, inexpensive test series can be developed to confirm or eliminate the suspected cause of equipment failure. As an example, hard-bluing can be used to verify the alignment and clearance of a gear set. This simple, inexpensive test requires very little effort and will absolutely confirm the wear pattern and meshing of the gear set. If alignment or excessive backlash is present, this test will confirm it.

When cavitation in a centrifugal pump is suspected, verify that at least one factor, such as suction leaks or low net positive suction head (NPSH), is present in the system. If not, cavitation cannot be the cause of failure.

Most of the causes contributing to problems that adversely affect plant performance can be grouped into one of several categories. Most equipment problems can be traced to misapplication, operating or maintenance practices and procedures, or simply age (not addressed in this module). Some of the other causes that are discussed include training, supervision, communications, human engineering, management systems, and quality control. These causes are the most common reasons for poor plant performance, accidents, and nonconformance to regulatory mandates. However, human error may contribute to, or be the sole reason for, the problem.

Figure 3–8 illustrates the most common causes of events that impact plant safety, environmental compliance, process performance, and equipment reliability.

Misapplication

Misapplication of critical process equipment is one of the most common causes of equipment-related problems. In some cases, the reason for misapplication is poor design, but more often it results from uncontrolled modifications or changes in the operating requirements of the machine.

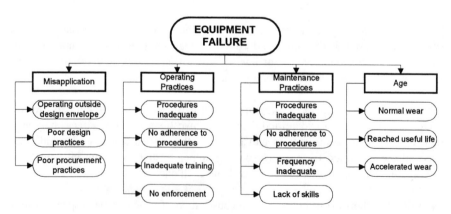

Figure 3–8 Common causes of equipment failure.

Poor Design Practices

Some equipment problems result from inadequate plant-engineering practices. Downsizing, a growing trend in U.S. industry, often diminishes or eliminates in-house design and plant-engineering capabilities. However, this can result in poor design practices, which have a major impact on equipment reliability. Even in those facilities where in-house design capability is not eliminated, the quality of procedures and practices can suffer such that it affects the design process and contributes to problems.

During the design-review phase of an investigation, the investigating team should have evaluated the design process to determine if it is adequate for the particular application. If not, the obvious corrective action is to modify the procedures and practices that are used for the design function.

Procurement Practices

The methods for procuring new and replacement equipment often contribute to equipment-related problems. The major factors contributing to procurement deficiencies are inadequate specifications, substituting vendors or machines, omitting life-cycle cost evaluations, and not obtaining vendor evaluations.

Inadequate Specifications

A major contributor in this category is the absence of adequate procurement specifications. Generally, specifications do not include enough detail to ensure that the correct part or machine is purchased. Too often, the specifications are limited to a model or part number. However, this type of specification does not define the specific application or range of operations that the machine must provide.

Substitutions

The defined role for the purchasing process is to limit the cost of new and replacement parts and machinery. As a result, procurement personnel often substitute parts or machines that they perceive to be equal. Unfortunately, the substitutes are not always exactly the same. The vendors themselves also contribute to this type of problem. They will offer their equivalent to the requested specification without reviewing the intended application. In some cases, this merely is an oversight. In others, it is an intentional effort to obtain the order, regardless of the impact on plant performance.

Low Bid versus Life-Cycle Cost

Procuring replacement or new equipment based strictly on *low bid* rather than *life-cycle cost* is another major contributor to equipment-reliability problems. In an effort to reduce costs, many plants have abandoned the process of vendor selection based on the total lifetime cost of equipment. This error continues to lower equipment reliability and may be the root cause of many problems.

Vendor Evaluations

The quality of new and rebuilt equipment has declined substantially over the past ten years. At the same time, many companies have abandoned the practice of regular inspections and vendor certification. The resulting decline in quality is another contributor to equipment malfunction or failure.

Poor Operating Practices and Procedures

Poor operating practices and procedures play a major role in equipment-reliability problems. Figure 3–9 illustrates the more common sources of problems caused by procedures. Problems generally result because procedures are not used, are inadequate, or are followed incorrectly. Any one of these categories may be the root cause, but it usually occurs in conjunction with one or more of the other causes discussed in this section. For example, *use not enforced* also is a supervision problem and the true root cause must include an evaluation. The key to evaluating a potential root cause is to determine if the problem is inadequate procedures or the failure to follow valid procedures.

Many SOPs used to operate critical plant production systems are out of date or inadequate. This often is a major contributor to reliability and equipment-related problems. If this is judged a potential contributor to the specific problem being investigated, refer to Figure 3–9 for the subclassifications that will help identify the root cause and appropriate corrective actions. Procedure problems have a more universal impact on reliability and performance in that there is an extremely high probability that the failure or problem will recur.

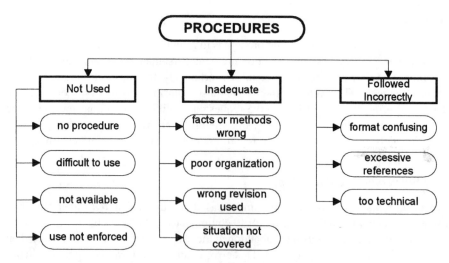

Figure 3–9   Common root causes of procedure-related problems.

Lack of enforcement is one reason procedures are not followed consistently. The real cause for this may be many factors. Training or inadequate employee skills commonly contribute to problems that affect plant performance and equipment reliability. The three major contributors to training-related problems include no training, inadequate training, and failure to learn.

The reasons underlying inadequate skills vary depending on the plant culture, workforce, and a variety of other issues. Figure 3–10 provides the common reasons for poor skill levels that are caused by training issues. In addition to the subcategories included in the figure, review the sections on Poor Operating Practices and Procedures and Supervision for problems where inadequate skill levels are potential contributors.

No Training

Many plants lack a formal training program that provides the minimal employee skills required to perform the procurement, operating, and maintenance tasks required to maintain an acceptable level of plant performance. In other cases, plant training programs are limited to mandated courses (e.g., OSHA, EPA, or other regulatory courses) that have little to do with the practical skills required to meet job requirements.

Inadequate Training

There are a variety of reasons why training programs fail to achieve a minimal employee skill level. Most are related to the procedures and methods used to develop and present the training courses. Some type of testing of the improvement in the skill level after the training is complete is the best way to determine the effectiveness of training programs.

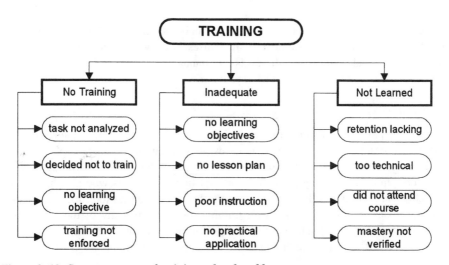

Figure 3–10 Common causes of training-related problems.

Failure to Learn

Employees fail to retain the instruction provided in training programs for two reasons: poor instruction and some failure by the employee. Poor instruction is a failure by the employer to provide the information required by the employee to perform a particular job in an easily understood format and at a pace that allows the employee to absorb the material. Failure by the employee to learn may be the result of their education level, fatigue, lack of desire, or lack of incentive.

Not all employees have the education and basic comprehension skills necessary to understand, master, and retain training to improve their basic operating and maintenance skills. In some cases, the employee may not have the ability to read and write at a level that permits minimal understanding of even the simplest training program. Employees that fall into this classification present a serious problem that is very difficult to correct. The only real solution is to provide remedial training courses focusing on basic reading, writing, and comprehension skills.

Fatigue is another factor that directly affects training. Many courses are taught outside of normal work hours and employees are expected to attend training classes after putting in a full workday. Unfortunately, this affects the students, ability to comprehend and retain the instruction.

Lack of motivation may be an employee attitude problem, but in many cases the real root cause is a failure in supervision and plant policies. If a lack of motivation is suspected, evaluate the potential causes included in the following sections on Supervision, Communications, and Management Systems before making a final decision on the root cause of the problem.

Supervision

Supervision includes all potential causes that can be associated with management procedures and practices. While most of the causes directly apply to first-line supervisors, they can also apply to all levels of management. Figure 3–11 provides the common factors of supervision-related problems.

Preparation

Lack of preparation is a common factor that contributes to, or is the sole reason for, plant-performance problems. In some cases, it is an employee problem with one or more supervisors or managers, but more often it is a failure in the overall plant-management philosophy. Many plants rely on first-line supervisors who also have other direct or indirect responsibilities. As a result, the supervisor has too little time to prepare job packages, provide hands-on training, or perform many of the other tasks essential for consistent plant (and worker) performance. If this is the case, citing poor supervision as the root cause of a problem is inappropriate and incorrect.

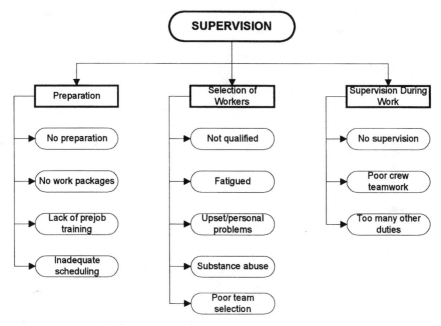

Figure 3–11 Common factors of supervision-related causes.

Another common reason for failure in this category is lack of supervisor training. Few first-line supervisors have been given the basic skills training that they need for their position. The materials presented previously under Poor Operating Practices and Procedures should be included in any evaluation where poor supervision is a potential source of the problem.

Worker Selection

Selection of workers assigned to specific jobs or tasks is a common contributor to problems. Generally, workers are assigned tasks based on the supervisor's perception of their abilities. In part, this is because skill levels are not universal across most plants and the supervisors attempt to put their best people on sensitive or critical jobs. Problems with this approach are that some members of the workforce become overloaded, which causes less skilled personnel to be used for job functions they are not qualified to perform. When evaluating potential root causes that may fall into this category, refer to the previous subsections on training problems.

Supervision During Work

Most of the problems that result from this cause result from the dual functions that many supervisors must provide. Due to other responsibilities, they are unable to directly supervise work as it is being done. A second potential cause of this problem is the lack of skill or training of the supervisor. Cross-check any potential problems that may be attributed to this category to ensure proper identification of the root cause.

Communications

Communication failure is another major category that contributes to plant problems. Figure 3–12 provides the typical causes for communications-related problems.

Any problem that may be attributed to one or more of these categories should be cross-referenced to the sections on Poor Operating Practices and Procedures and Management Systems to ensure the correct cause is identified.

Human Engineering

Deficiencies in this category are not as visible as some of the others, but they contribute to many plant problems. These causes, which are shown in Figure 3–13, are primarily the result of poor workplace design and errors in work flow.

Worker-Machine Interface

The layout of instrument panels, controls, workstations, and other equipment has a direct impact on plant performance. For example, when the workstation design forces an employee to constantly twist, bend, or stoop, the potential for injury, error due to fatigue, or loss of product quality is substantially increased. A properly designed workstation can help eliminate potential problems.

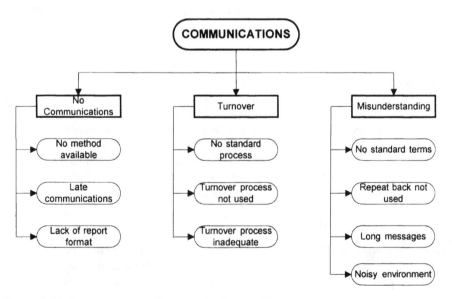

Figure 3–12 Common causes of communications problems.

Work Environment

The work environment has a direct impact on the workforce and equipment reliability. Improper lighting, lack of temperature control, and poor housekeeping are the dominant failure modes within this classification. In some cases, a poor environment is a symptom rather than the root cause of a problem. For example, supervision, procedures, or management system problems may be the real cause. Make sure that the evaluation is detailed enough to isolate the true root cause.

Complex Systems

Production systems are becoming more and more complex. As a result, the skill levels of operators and maintenance personnel also must improve to achieve and maintain acceptable reliability levels. While the categories included in this section may be perceived as a viable root cause of a specific investigation, most are a combination of the system's complexity and other issues, such as training, employee motivation, and others.

Management Systems

Figure 3–14 illustrates the common root causes of management-system problems: policies and procedures, standards not used, and employee relations. Most of these potential root causes deal with plant culture and management philosophy. While hard to isolate, the categories that fall within this group of causes contribute to many of the problems that will be investigated.

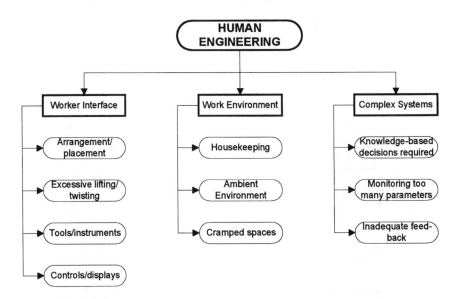

Figure 3–13 Common causes of human engineering problems.

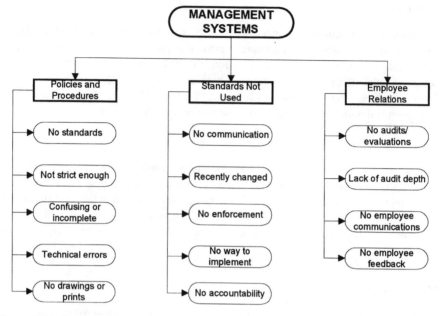

Figure 3–14 Common causes of management systems problems.

Quality Control

Quality control is a potential root cause primarily in process-deviation problems. While such problems may contribute to equipment failure or other types of problems, their dominant role is in quality and capacity-related investigations. Figure 3–15 lists the more common quality-control causes created by no or inadequate inspection.

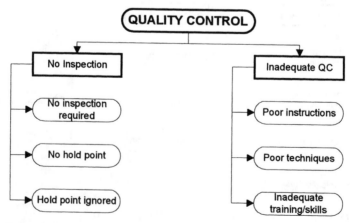

Figure 3–15 Quality control-related causes.

EVALUATING POTENTIAL CORRECTIVE ACTIONS

The RCFA process defines one or more potential corrective actions that should resolve the incident or event under investigation. The next step in the analysis is to determine which, if any, of these corrective actions should be implemented. Figure 3–16 illustrates a simplified decision tree that defines multiple potential corrective actions. These actions include a wide range of possibilities from doing nothing to replacing the machine or system.

Which corrective action should be taken? Not all actions are financially justifiable. In some cases, the impact of the incident or event is lower than the cost of the corrective action. In these cases, the RCFA should document the incident for future reference, but recommend that no corrective action be taken.

The basic machine knowledge provided in Part Two of this book and troubleshooting guides in Part Three provide the guidance needed to select the potential corrective actions required to resolve the failure and to prevent a recurrence.

The objective of this task is to define a short list of corrective actions that will resolve the root cause of the failure. At this point, cost is no factor and the list should include the best solutions for the identified cause. For example, a pump failure caused by cavitation can be corrected by (1) eliminating air or gas entrainment, increasing the

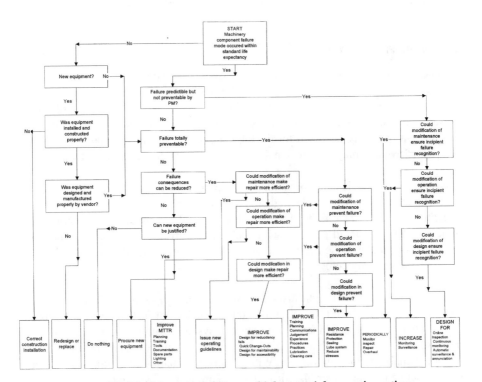

Figure 3–16 Simplified decision tree defining multiple potential corrective actions.

actual net positive suction head (NPSHA), lowering suction liquid temperature, or increasing suction pressure. Depending on the type of cavitation, one or any combination of these corrective actions will eliminate the problem.

Once the short list has been generated, however, cost must be considered. Each potential corrective action should be carefully scrutinized to determine if it actually will correct the problem. *This is important because the normal tendency in less formal evaluations is to fix the symptoms of problems rather than the true root cause.* Therefore, this phase of the evaluation should evaluate each potential corrective action to ensure that it eliminates the real problem.

However, on some occasions, correcting the symptoms or implementing a temporary solution is the only financially justifiable course of action. In these instances, the recommendation should clearly define the approach that should be taken. It should outline the rationale for the decision and describe the limitations or restrictions that the partial correction will have on plant performance, equipment reliability, and other factors of plant operation and maintenance.

Although the number of temporary fixes should be limited as much as possible, they often are unavoidable. Every effort should be made to implement *permanent* corrective actions, and the goal is to eliminate all negative factors associated with the event or incident. While there generally is a corrective action that meets this goal, the challenge is to find the best acceptable action that also is cost effective.

COST-BENEFIT ANALYSIS

A full cost-benefit analysis is the final step before recommending a course of action. It is needed to compare costs with the benefits derived from the corrective action(s) being considered. For example, many machine failures are the direct result of misapplication. In these instances, the "best" solution would be to replace the machine with one that was properly designed for the application. However, this solution usually is very expensive and may require a major modification of the installed system.

A cost-benefit analysis is simply a direct comparison of the actual total *costs* associated with an activity (e.g., replacing a pump or modifying a production line) with the *benefits* to be derived from the change. The amount of historical data available determines the level of detail possible with such an analysis. If sufficient information is available, the analysis should include all costs associated with the area, machine, or system being investigated.

While this may sound straightforward, developing an *accurate* cost-benefit analysis sometimes is very difficult. In part, this is due to the lack of factual, detailed historical data for many of the machines and production systems within the plant. As a result, much of the cost data, such as power consumption, repair costs, and repair intervals, are not available to the investigator. Without this information, it is almost impossible to fully define the current or historical cost of a production system or area.

The analysis should include the effect of the problem on downstream production facilities. Since most plants are integrated to some degree, a problem in one area of the plant generally has a direct impact on all downstream plant areas. For example, if the preparation area fails to deliver an adequate volume of prime-quality flake in a timely manner, the performance of all downstream production areas is adversely affected. Therefore, it is important that the loss of revenue, potential increase in conversion costs, and resultant benefits of the corrective action on the downstream areas be considered in the cost-benefit analysis.

Cost Analysis

The cost analysis consists of two parts. The *first* part should quantify the impact of the problem, incident, or event on the process. For this analysis, the impact must be defined in financial terms rather than in terms of delays, downtime, and other traditional tracking mechanisms. If the problem is proven to cause an increase in unscheduled delays, it must be defined in costs rather than hours or percentages. The *second* part of the analysis should define all costs directly or indirectly associated with actually implementing the recommended corrective action(s).

Process-Related Costs

The Cost-Accounting Department in most plants can provide some assistance converting non-financial data into actual costs. They should have established guidelines that define the operating costs for each production unit or system within the plant in terms of dollars per unit time (e.g., dollars/hour). In most cases, these costs consist of production labor, consumable materials, power consumption, and plant overhead.

While operating costs are the "true" costs, they do not include all the financial impacts that result from delays or downtime. Other process-related costs that must be included are capacity losses and delivery-schedule slippage.

Materials Most problems that result in equipment failure or deviations in process performance have a history of abnormal material costs, which should be quantified as part of the cost-benefit analysis. These costs include replacement parts, spare machinery, special expediting charges, and all other material-related costs incurred by the machine or process system being investigated.

Note that the cost analysis should define only those costs that are unusual and are incremental costs due to the specific problem. The best way to determine the actual incremental material costs due to the abnormal machine is to compare the total incurred costs to the recommended material cost provided in the vendor's O&M manual. In these manuals, the vendor provides specific recommendations on the number of spares, replacement parts, and expendables that should be required by the machine. This data provide the means to *calculate* the total material cost that should have been incurred for the investigated machine. Incremental cost is the difference between the actual cost incurred and the projected cost derived from vendor information.

Labor Incremental labor costs should include all unusual costs incurred to operate, maintain, or repair the machine or system being investigated. These costs should include overtime premium, contract labor, outside vendor or shop support, and any other costs above the normal operating and maintenance budget that were incurred as a result of the abnormal behavior of the machine or system.

Capacity Loss Many equipment and other process-related problems result in decreased product output, or capacity. In the cost analysis, capacity losses should be defined both in terms of production units (i.e., tons, pounds, etc.) and actual revenue losses (i.e., dollars). Using both of these measures provides a true picture of the situation.

An accurate estimate of the total downtime, including testing following the repair and reduced-capacity startups, is essential for this phase of the evaluation. However, the normal tendency is to estimate the actual time required to repair, replace, or modify the system but to omit the inherent problems that will be encountered during the startup process.

Startups following any major activity, such as a repair, upgrade, or modification, rarely are trouble free. Production capacity almost always is lost while "glitches" are corrected. *In some cases, the losses caused by startup problems are significantly greater than those incurred simply from the maintenance activity.* There are two viable ways to determine this cost.

The first is to use the plant's business plan to obtain the *planned* production rate for the area or system under investigation. Most business plans define a rated or planned capacity, usually expressed in terms of production units per time (i.e., pounds/hour). The logic used to develop these capacity rates vary from plant to plant, but typically they are a *weighted* rate based on demand, seasonal variations in business, and historical data. In some cases, the rate will be substantially below the actual design capacity of the system.

The second method is to use the design-capacity rates defined in the design review. This approach is more accurate since it is based on actual design limitations. The functional specifications of each production system within the plant define minimum, mean, and maximum capacity rates. In most cases, the mean or average capacity rate is used when developing the cost analysis. This approach is conservative and easily can be justified.

Delivery-Schedule Slippage Although capacity loss is a critical element because of its impact on cash flow and profitability, another important consideration is the delivery schedule. Customers expect on-time deliveries and may cancel orders if schedules are not met, adding further impact to the cash flow and profitability. Therefore, extreme caution should be exercised if the downtime required to implement a potential corrective action has a significant effect on delivery schedules.

Before acting on a recommendation for corrective action that seriously affects delivery, the decision to implement and its timing should be closely coordinated with the Sales and Customer Service Departments. These groups are in a better position to evaluate the potential impact of product delays than the investigator or investigating team.

Implementation Costs

Implementing change, no matter what the change is, will have a one-time cost. Most important, implementing many potential corrective actions will force a system shutdown. In many cases, this single factor will determine which—if any—of the potential corrective actions can be implemented.

Therefore, the loss of production capacity and the revenue it generates is a cost that must be considered in the cost-benefit analysis. The analysis must include all the costs, direct and indirect, required to implement the recommended change. For example, replacing a damaged or defective pump will incur costs that include lost capacity because of downtime, pump procurement, labor to install the new pump, and other miscellaneous costs. Other corrective actions, such as improving the skill level of plant personnel, also have implementation costs. In this case, costs include the salary of the employees to be trained, training course development costs, instructor cost, and so forth.

Many of these costs are obvious (i.e., material and labor required to install a machine), but others are hidden and easily overlooked. For example, replacing a machine or system component normally will require some period of downtime to make the change. The losses in capacity and labor costs for the idle product crew not always are apparent and can mistakenly be omitted from the analysis.

Materials Many of the potential actions discovered through RCFA involve repairing, replacing, or modifying existing machinery or systems. Each action must be evaluated to determine the real material costs of the proposed changes.

Exercise great care to identify *all* significant material costs. For example, replacing a centrifugal pump may require a new foundation, complete repiping of both the suction and discharge system, additional spare parts, and a myriad of other material-related costs. It is crucial that the evaluation accurately identify all direct and indirect costs due to the potential corrective action.

Labor Estimating labor costs typically is more difficult than determining material costs. Without accurate records of the level of effort required to repair, replace, or modify a machine or process, it is extremely difficult to estimate the actual labor involved in a potential corrective action. Simple tasks, such as replacing a bearing, often can be quantified—and the time involved may *appear* to be relatively insignificant. However, many machines require complete or partial disassembly before the bearing can be replaced.

Nevertheless, try to be as accurate in the labor estimate as possible. Solicit the advice of plant engineers, maintenance personnel, and vendors. These individuals can provide valuable assistance in this phase of the evaluation.

In addition, do not forget hidden costs. Replacing or modifying an existing machine or system involves labor costs for the internal engineering and procurement activities. While these levels of effort are determined by the complexity of the activity, they can be substantial.

Benefits Analysis

The benefits analysis involves defining the benefits derived from implementing specific corrective actions. When doing a benefits analysis, however, the tendency is to overestimate the significance of the benefits of a modification or change in a critical process system. In RCFA, the objective is to *quantify* the actual improvement that will be derived from the recommended corrective action to ensure that the potential benefits are real and significant. Benefits generally can be quantified as improvements in process-related costs, which result in reductions in cost per unit, increased revenue generation due to higher capacity, and cost avoidance of chronic high maintenance costs.

Improvement in Process-Related Costs

The format of the benefits analysis should mirror the cost categories (e.g., material and labor costs) so that a comparison can be made. If the recommended corrective action is valid, there should be a measurable improvement in process-related costs.

Reduction in Unit Cost

One potential benefit is a reduction in the total production and maintenance costs per unit (i.e., pound or ton). For example, when a machine with a history of chronic reliability problems is replaced, normally a measurable increase in production capacity will occur. In some cases, this improvement may occur simply because the capacity of the replacement machine is greater than the one replaced. Regardless of the reason, the increase in capacity should reduce the total cost per unit produced simply because greater volume is produced.

The benefits analysis should establish a reasonable level of improvement in this cost category. The simplest method is to define the rated capacity of the production system under investigation. For most production systems, the percentage of operating capacity expected to be used can be found in the business plan.

The cost analysis should have quantified the losses (i.e., higher cost per unit produced) without corrective action. The benefits analysis should quantify the gain (i.e., reduction in cost per unit produced) to be achieved after correction. The mathematics are relatively simple. If the system was operating at a level equal to 60 percent capacity before correction and 90 percent afterward, the differential of 30 percent has a quantifiable value. The value of each unit of product produced (either in-process or finished goods) multiplied by the capacity gain (i.e., 30 percent) is a quantifiable benefit.

In addition to the capacity gain, the increase in availability has the additional benefit of reducing the cost per unit of both production and maintenance. Labor costs make up about 60 percent of the costs required to produce and maintain process systems and are relatively fixed. With a production capacity of 100,000 units and total production costs of $100,000, the cost per unit is $1. If the capacity increases to 200,000 units, the cost drops to $0.50 per unit.

Increased Revenue

Increased capacity, as illustrated in the preceding section, is a major benefit that may be derived from implementing corrective actions. In addition to the reduction in unit costs, this increase also will provide additional revenue for the plant, assuming there is a market for the additional product that can be produced. If the sales value is $100 per unit and the 30 percent increase represents 30,000 units, the benefit is $3 million per year.

Remember that the potential benefit of the improvement is over the useful lifetime of the process system being improved. For a typical machine, the life generally is 20 years. Therefore, the lifetime benefit is 20 times $3 million or $60 million.

Cost Avoidance

The second type of benefit that should be considered is cost avoidance or the eliminating unnecessary or excessive costs, such as high maintenance costs created by a machine with a history of chronic problems. To establish this type of benefit, the investigator needs to gather the cost history of the machine or system that is being investigated. These data provide the reference or existing costs.

The second step is to calculate the projected costs, in this case maintenance, of the upgraded or new machine. The simplest way to develop these future costs is to use the vendor's recommendations for routine maintenance and upkeep. Using these recommendations and the internal labor and material costs, the annual and lifetime costs of the upgrade or modification can be calculated.

Cost avoidance should include *all* unnecessary or avoidable costs that have been incurred as a result of the problem being investigated. The following are examples of avoidable costs:

1. Losses incurred due to poor quality (e.g., scraps, rejects, and reworks).
2. Overtime premiums for production and maintenance labor.
3. Expedited vendor deliveries or outside repair work for emergency shipments and repairs.

4. Capacity losses due to poor equipment condition, improper operation, inadequate maintenance, and the like.
5. Fines and penalties caused by spills, releases, or other nonconformance to regulatory requirements.

Cost-Benefit Comparison

Once the costs and the projected benefits have been quantified, the final step is to compare them to determine the value (either positive or negative) of the recommended improvement. If the recommended change is valid, the value of the benefits should far outweigh the costs.

The actual differential required to justify a modification or upgrade to management varies from company to company. Most require a payback period of three years maximum, but some require a one-year payback. Regardless of the time line, the projected benefits derived as part of the cost-benefit analysis must clearly show that the recommended corrective action will offset all incurred costs *and* generate a measurable improvement in one or more of the cost categories.

As a general guideline, the cost history and projected savings or gains should include a three- to five-year time period. In other words, the cost portion should include three to five years of historical costs, and the benefits should be projected over an equal period. This method provides a more accurate picture of the real improvement that will result from the recommended change.

REPORT AND RECOMMENDATIONS

The next task required in the RCFA is a report that includes a complete description of the incident or event, identification of the specific cause(s), and recommendations for its correction. The report should include the following information:

- Incident summary,
- Initial plant condition,
- Initiating event,
- Incident description,
- Immediate corrective actions,
- Causes and long-term corrective actions,
- External reports filed,
- Lessons learned,
- References and attachments,
- Investigator or investigating team description,
- Review and approval team description, and
- Distribution list.

Incident Summary

The incident summary should be a short, concise description of the incident or event. It should not elaborate on the factors that may have contributed to the incident.

Initial Plant Condition

This section should include a brief description that defines the plant's status at the start of the incident. It should include any abnormal conditions that contributed to the incident. This section is not intended to provide a quantitative analysis of the incident, but should be limited to a clear description of the boundary conditions that existed at the time of the event.

Event Initiating Investigation

Give a brief description of the initial failure or action that triggered the incident or led to its discovery and the resulting investigation. Do not use specific employee names or titles in this section of the report. Instead use the codes and descriptors that identify functions within the affected area.

Incident Description

This section should include a detailed chronology of the incident. The chronology should be referenced to the sequence-of-events diagram developed as part of the analysis. The diagram should be included as an appendix to the report.

This section should include a description of how the incident was discovered, the facts that bound the incident, identification by component number and name of any failed equipment, safety-system performance, control-system actions, significant operator actions and intervention, and transient data for important plant parameters.

It also should include any special considerations observed in the incident, such as unexplained or unexpected behavior of equipment or people, inadequate or degraded equipment performance, significant misunderstandings by operations or maintenance personnel, common failure modes, progression of the event beyond the designed operating envelope, violation of technical specifications or design limits, or failure of previously recommended corrective actions.

Immediate Corrective Actions

Many of the failures or events having a direct impact on production require immediate corrective actions that will minimize downtime. As a result, temporary actions often are required to permit resumption of production. This section should describe what intermediate or *quick-fix* actions were taken to permit resumption of production.

Causes and Long-Term Corrective Actions

This section of the report should clearly describe the specific root cause(s) that triggered the incident. It should detail specific recommendations for long-term corrective actions.

Root Cause(s)

The root cause(s) description should be complete and provide enough detail and support data that the reader can fully understand and accept the rationale used to isolate the root cause(s). This section should include all contributing factors, such as training and supervision, that contributed to the incident.

Recommended Long-Term Corrective Actions

This section should provide a clear, complete description of the corrective actions required to prevent recurrence of the incident or event. It should incorporate conclusions derived from the cost-benefit analysis developed previously. The full analysis should be included as part of the submittal package forwarded to management for approval.

External Telephone Calls Made and Reports Filed

This section should document any external telephone calls placed and reports written and filed. These requirements are summarized in Table 5–1 found in Chapter 5.

If the incident involves regulatory-compliance issues, telephone calls may have been placed and reports filed to the local police and emergency response team, the Coast Guard's National Response Center (NRC), the state's Emergency Response Commission, the local Emergency Planning Committee, the U.S. Environmental Protection Agency (EPA) regional office, the U.S. Department of Transportation, or the state's water-pollution control agency.

Lessons Learned

This section should describe the lessons learned that should be passed on to the appropriate personnel. This information should be disseminated through formal training or some other means, such as individual feedback or required reading. This section also should designate the person (by name) responsible for ensuring that the lessons learned are communicated, a completion date for the communication, and a list of the specific plant personnel who should receive the communications.

References and Attachments

List all references and attachments used as part of the investigation. Copies or excerpts of appropriate procedures, logs, computer printouts, instrument charts, and statements of involved personnel should be included as appendices.

Investigator or Investigating Team Description

This section should include a list and descriptions of the personnel involved in the investigation.

Review and Approval Team Description

RCFA reports always should be reviewed and approved before being distributed to plant personnel. The approval chain will vary from plant to plant but should be established before implementing a RCFA. The review process is important in that it will reduce the potential for errors or misinterpretation that is an inherent part of an investigation process.

Distribution

Include a list of the plant personnel, including appropriate contractors, who should receive a copy of the final report.

VERIFY CORRECTIVE ACTION

Once the corrective action has been approved and implemented, the final task required for a thorough RCFA is to verify that the corrective action actually fixed the problem. After the recommended corrective action has been implemented, a series of confirmation tests should be conducted to ensure reliable operation of the corrected application or machine.

For most common plant machinery, a well-planned, multichannel real-time vibration analysis should confirm the corrective action. A series of vibration measurements can be developed and implemented to measure the operating dynamics of the new or modified installation. The series should include the full operating envelope of the system, including any changes in operating practices that were part of the recommendation. The results of this test series will confirm the validity of the corrections and provide assurance that the modified system operates reliably.

4

SAFETY-RELATED ISSUES

An incident or event that results in injury or death must be fully investigated using RCFA techniques. Figure 4–1 is a logic tree for investigating such an accident or injury. Refer to Chapter 5 for a detailed discussion of Occupational Safety and Health Administration requirements for such an investigation.

Because of potential liability and to facilitate the investigation, the accident scene *must* be isolated and preserved until the investigation is complete. In the section on preservation of evidence in Chapter 3, preserving the scene was optional, but in the case of an accident it is required. The entire area surrounding the accident should be locked-off and barriers erected to prevent access by unauthorized personnel.

Once secured, the scene must be fully documented. Photographs, sketches, and other documentation are needed to "freeze" all the parameters that may have directly or indirectly contributed to the accident. These data must be gathered for analysis at a later time.

The majority of events or incidents, excluding catastrophic equipment failure, can be directly traced to one or more of the generic root cause classifications illustrated in Figures 3–8 through 3–15. The investigation process first will define which of these major classifications contributed to the event, then it will isolate the specific root cause(s) that resulted in the incident. The logic used to investigate accidents or events that resulted in, or could have caused, personal injury is shown in Figure 4–1. This process identifies those issues most likely to have contributed to this type of incident.

In those safety-related events that also result in equipment failure, these steps should be used in conjunction with those outlined in the section on Equipment Troubleshooting in Chapter 3. As in all RCFA, the first step is classifying the incident or event. The steps outlined in the section on Failure Analysis or Cause and Effect Analysis should be used for this task. Clear concise answers to the questions in the section on Problem Clarification will provide a clear definition of the event.

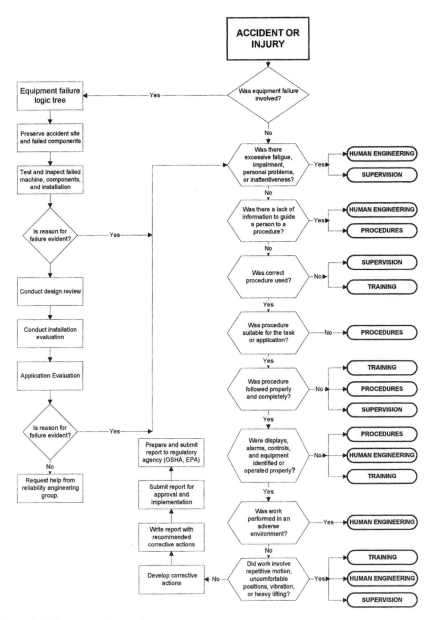

Figure 4–1 Logic tree illustration.

The primary tool for this initial task is the interview process defined in Chapter 3. The process should use the two-on-one interview technique. Because of potential liability claims that may result from any incident involving personnel injury or death, extreme care must be taken to ensure accuracy of the acquired data and eliminate any hint of misinterpretation or prejudice on the part of the interviewer. The use of two interviewers will help avoid problems later.

In addition to the questions provided in the section on Identifying Root Cause, accident or potential safety problems need additional clarification of several factors: physical impairment, personal problems, and repetition. Many accidents are the direct result of one or more of these factors. Therefore, the interview process must be configured to address these potential forcing functions and every effort made to determine if they contributed to the accident.

The investigator should use caution and empathy when addressing these issues with plant personnel. Many will be reluctant to discuss them with their peers, and even more so with an investigator who may be perceived as one who would use the information to punish the employee.

FATIGUE

Work records should be checked to verify the total hours, including overtime, that the employee had worked prior to the incident. This review should include a period of at least six weeks prior to the incident.

However, on-site work records may not always provide a clear picture of the employee's fatigue, as some employees work outside the plant. This work may be a second job or could be nonpaid work such as farming, building or remodeling a house, attending school, or other activities that limit the amount of rest that the individual had prior to the accident.

PHYSICAL IMPAIRMENT

Without the employee's permission, the investigator cannot verify his or her physical condition. Unless the company's policy specifically grants it the right to check medical records both internally and outside the facility to verify the worker's physical condition, the only way this factor can be judged is on information provided by the employee.

If the employee had a prior accident, there should be a record of any treatment and work release, which usually is required before he or she can return to full-duty status. These records should be available to the investigating team. If the employee was on restricted duty, the investigator also should determine if the restriction had any bearing on the incident.

PERSONAL PROBLEMS

Distractions caused by personal problems are a leading cause of work-related accidents. However, extreme caution must be exercised when trying to ascertain if it was a contributing factor to the incident. If the employee does not respond to a question, the interviewer should not pursue that subject further.

An employee cannot be forced to discuss his or her personal life. If the employee does not voluntarily provide the information, the interviewer should seek guidance from the Human Resources manager or the company's legal advisor.

REPETITION

Repetitive tasks often contribute to accidents. Jobs that require constant repetitive motion or movement may lead to boredom, inattentiveness, and premature fatigue. Therefore, the investigator should evaluate the functional requirements of the employee involved in the accident or safety violation to determine if repetition was a contributing factor in the incident.

5

REGULATORY COMPLIANCE ISSUES

RCFA is a viable tool that can be used to investigate regulatory compliance problems that arise in a plant. The logic and decision trees for this type of investigation are similar to those used for equipment failure and safety investigations but must be modified to meet both the nature of the incident and the specific requirements of the regulatory body.

Regulatory compliance problems associated with work-related incidents generally fall into the domain of safety, which is regulated by the Occupational Safety and Health Agency (OSHA), and environment, which is regulated by the Environmental Protection Agency (EPA) and state (although a state may defer to the EPA through a formal agreement) and local governments. This section provides guidelines for investigating incidents that fall under the jurisdiction of regulations by these groups.

NOTE: The material presented in this section does not constitute legal advice and is provided for informational purposes only. Please consult the appropriate agencies directly and seek qualified legal council for this type of assistance.

SPILL HAZARDS

Be certain that spill hazards are reviewed as part of the plant's internal audit program. Audit-team members should be trained to look for deficient containers, fall hazards that can contribute to a spill, improper container labeling, and other potential hazards. Major causes of chemical spills are

- Mishandling of equipment and supplies in the work area (e.g., use of improper containers),
- Mishandling during transportation (even within a facility),
- Leaks from process equipment such as flanges or valves, and
- Improperly stored containers (e.g., storage in a damp location leads to rusty, deteriorating containers).

WORKER EDUCATION AND COMMUNICATION

Employers must ensure that their workers know the potential hazards of the chemicals they work with, how to protect themselves against those hazards (e.g., safe practices, personal protection equipment), and what to do in case of an emergency. Accordingly, OSHA has established basic communication requirements under the *Hazard Communication Standard* to inform workers about chemicals in use in the workplace. Under this standard, chemical makers must meet the following requirements:

- Manufacturers must determine the physical and health hazards of each chemical they produce.
- Users must be informed of the hazards through container labels and Material Safety Data Sheets (MSDSs).
- A written hazard-communication program must describe the employer's efforts to tell employees about the standard and how it is being implemented at the work site.

PREVENTION

Preventive measures include everything from safe work practices to training programs, good housekeeping, regular audits, administrative and engineering controls, and chemical-protection equipment.

Workers must be properly trained to handle the responsibilities assigned to them, with members of a designated spill team undergoing more specialized training than the average employee. All worker training should include the basic instruction that *employees not involved in cleanup must stay away from a spill.* This is important because of risks such as inhaling chemical vapors or fire or explosion hazards. Another important lesson to teach workers is how to prevent a spill or leak from becoming worse. Proper containment can mean the difference between a small problem and a huge, dangerous mess.

RECOMMENDED SPILL RESPONSES

The following steps are recommended responses when a chemical spill occurs.

1. If necessary, evacuate the area of untrained spill-response personnel.
2. Check labels, MSDSs, and other key documentation to identify what has spilled.
3. Assign at least two qualified individuals to the cleanup. They may be internal staff or from an outside firm that you have prearranged to handle this task. Two workers are needed in case one is overcome or injured.
4. Promptly clear the spill area by ventilation.
5. If the spilled material is on fire, douse the flames in a safe manner.

6. Limit the spread of the spilled substance by containing it with a dike. Use an absorbent material appropriate for the type of spill (e.g., polymer-type absorbents such as spill booms, granular materials such as kitty litter or ground corncobs).

However, when absorbent materials are used, they become as hazardous as the material absorbed. Therefore, federal, state, and local regulations must be consulted before disposing of these contaminated materials. In some cases, a neutralizer can be used to convert one material into another (e.g., an acid neutralized with a base produces salt and water).

CAUTION: Specialized worker training legally is required to use absorbent materials on spills because of the chemical reactions that can occur. For example, organic materials should never be used on nitric acid spills because a fire will result!

WORKER RESPONSIBILITIES

Employees must take most of the responsibility for protecting themselves from chemical hazards. However, adequate training and frequent reminders from the employer can help ensure that they take that responsibility. The following are some basic chemical-safety tips to incorporate in a worker-safety training program or to post on the bulletin board:

- Pay attention to the training that is provided.
- Know what hazardous chemicals are used in your work area.
- Read labels and MSDSs before starting a job.
- Do not use chemicals from unlabeled containers.
- Follow manufacturers' instructions for chemicals and equipment.
- Follow company rules and procedures and avoid shortcuts.
- Wear all personal protection equipment (PPE) required by your organization and ask your supervisor if you are not sure.
- Keep containers closed when not in use.
- Check containers regularly for leaks.
- Make sure that equipment is in good working condition before use.
- Keep incompatible materials apart.
- Keep flammable and explosive materials away from heat sources.
- Make sure the work area is adequately ventilated.
- Do not bring food or drinks into a work area where chemicals are used.
- Wash before eating or drinking.
- Cleanse tools, equipment, and clothing that have been exposed to hazardous materials before storing or reusing them.
- Dispose of contaminated materials properly.
- Ask a supervisor what to do with old or unused chemicals in your work area.

LEGISLATION AND REPORTING REQUIREMENTS

In any industrial facility, from offices to factories and laboratories, spills happen and create a variety of risks to workers. Inside a plant, spills result in chemicals on the floor, in the air, or on the workers themselves. When releases occur outside the plant (e.g., chemical releases from tank cars or trucks, the spread of noxious fumes from an internal spill), the potential for harm extends far beyond the facility, particularly with major catastrophes.

Catastrophes, such as the Bhopal chemical release, the *Exxon Valdez* oil spill, New York's Love Canal, and dioxin-contaminated Times Beach in Missouri, have led several federal departments and agencies to enact protective regulations. These protections are aimed at protecting a much broader range of people, property, and the environment than most regulations administered by OSHA.

Spills are covered by a variety of federal, state, and local reporting requirements; and substantial penalties can result to a company and its employees for failing to report certain spills. Initial release notification usually is required immediately or within 24 hours of the release, and in some cases, written follow-up reports are required. Some of the applicable legislation is listed next and Table 5–1 lists some of the major reporting requirements for chemical spills that are specified by these acts.

- OSHA's Process Safety Management (PSM) Standard.
- OSHA's Hazardous Waste Operations and Emergency Response (HAZWOPER).
- Superfund: Comprehensive Environmental Response, Compensation, and Liability Act (CERCLA).
- Superfund Amendments and Reauthorization Act (SARA): SARA Title II contains the Emergency Planning and Community Right-to-Know Act (EPCRA).
- Resource Conservation and Recovery Act (RCRA).
- Toxic Substances Control Act (TSCA).
- Clean Water Act (CWA).
- Department of Transportation (DOT) rules for packing and shipping.
- Hazardous Materials Transportation Act (HMTA).

OSHA Process Safety Management Standard

OSHA legislation focuses primarily on individual workplaces and is intended to prevent explosions, spills, and other disasters. The Process Safety Management (PSM) standard covers large-scale makers and users of highly hazardous chemicals and other chemical manufacturers. However, small companies whose core business has nothing to do with chemicals also are vulnerable to spills (e.g., cleaning products, toner for the copying machine). Although the PSM standard does not specifically apply to smaller chemical spills, its principles still are valid. Note that RCRA regulations apply to these small spills.

Table 5–1 Major Regulatory Reporting Requirements for Chemical Spills

Regulation	Reference	Reporting Requirements
Superfund	40 CFR 302.4	Immediately report to the Coast Guard's National Response Center (NRC) the release of CERCLA hazardous substances in quantities equal to or greater than its reportable quantity.
EPCRA	40 CFR 355	Report releases of a reportable quantity of a hazardous substance to the state Emergency Response Commission (SERC) for each state likely to be affected. Also provide notice to the local Emergency Planning Committee (LEPC) for any area affected by the release.
RCRA	40 CFR 240-281	Notification to the NRC is required for releases equal to or greater than the reportable quantity of a RCRA hazardous waste. If the waste also is on the CERCLA list, that reportable quantity applies. If not, the reportable quantity is 100 pounds if the waste is ignitable, corrosive, reactive, or toxic.
TSCA	40 CFR 761.120 et seq. Section 8(e)	Immediately report by telephone to the EPA regional office any spill of a hazardous chemical that "seriously threatens humans with cancer, birth defects, mutation, death, or serious prolonged incapacitation, or seriously threatens the environment with large-scale or ecologically significant population destruction." A written follow-up report is required within 15 days.
CWA	Oil: 40 CFR 110-114 Haz: 40 CFR 116-117	Report any *oil spill* that occurs into navigable waters or adjoining shorelines to your regional EPA office and state water pollution-control agency if it violates water-quality standards, causes a sheen or discoloration of the water or shoreline, or causes a sludge or emulsion to be deposited beneath the surface of the water or on the shoreline. Immediate notification also is required to the NRC for the release of a *designated hazardous substance* in a reportable quantity during a 24-hour period if the spill is in or alongside navigable waters.

Table 5–1 Major Regulatory Reporting Requirements for Chemical Spills (continued)

Regulation	Reference	Reporting Requirements
HMTA		Generally, the transporter of hazardous materials (including wastes) must immediately report to the NRC and the state response center a release during transport if the release meets any of the following criteria: causes death or serious injury, involves more than $50,000 in property damage, involves the release of radioactive materials or etiological agents, requires public evacuation lasting at least one hour, closes one or more "major transportation artery or facility" for at least one hour, alters the flight pattern or routine of an aircraft. Even if none of these criteria are met, it should be reported if the carrier believes a spill or incident poses "such a danger" that it should be reported. Follow-up written reports are due within 30 days. The carrier also must file DOT Form 5800.1.

For all releases:

Always call 911 first to assure that first responders are dispatched to the scene to stabilize the release, render first aid, establish a perimeter, and extinguish/minimize the threat of fire or explosion.

Useful telephone numbers:

Coast Guard's National Response Center (NRC): 800-424-8802

EPA's national database of all toxic chemical release information: 800-638-8480

EPA's Emergency Planning and Community Right-to-know Information Hotline: 800-535-0202

CFR   = Code of Federal Regulations,
CWA   = Clean Water Act,
EPCRA = Emergency Planning and Community Right-to-Know Act,
TSCA   = Toxic Substance Control Act,
RCRA   = Resource Conservation and Recovery Act,
HMTA   = Hazardous Materials Transportation Act Source: Adapted by Integrated Systems, Inc., from *Environmental Compliance National Edition* (Issue No. 181, Business & Legal Reports, Inc., Madison, CT, Dec. 1996) and other sources.

This section summarizes the requirements for OSHA's *PSM of Highly Hazardous Chemicals, Explosives, and Blasting Agents* procedures for incident investigation (CFR 1910, Part 11.9, Section m).

Incident Investigation Requirements

The regulation states:

> The employer shall investigate each incident which results in, or could reasonably have resulted in, a catastrophic release of highly hazardous chemicals into the workplace.

To meet this requirement, a company must define an incident in terms specific to its facility. This includes an operational definition that indicates the number of pounds of the substance used in a particular process that would qualify as a "catastrophic" event.

Defining an incident in site-specific terms also includes defining the term *could reasonably have resulted in*. Appendix C of the regulation provides guidelines for clarifying this point. It includes definitions of *near misses* in which a catastrophic failure occurred but a chemical release did not occur. Clear guidelines should be established that provide the employee with a quantifiable means of defining those incidents that require a violation report.

Table 5–2 provides examples of hazardous chemicals that require investigation when a catastrophic release occurs or when one could have happened. These OSHA guidelines should be used in conjunction with site-specific procedures. For a complete listing of the reportable chemical used in your plant, refer to the site Hazardous Materials Policy and Procedure Manual.

Required Scope

OSHA 1910.119 does not mandate the specific type of investigation a plant must conduct when a reportable incident occurs. However, it provides stipulations that must be met for the following: investigator qualifications, time requirements, report content, review process, and corrective actions.

Table 5–2 Examples of OSHA-Listed Chemicals

Chemical Name	Chemical Abstract Service	Threshold Quantity (pounds)
Ammonia, Anhydrous	7664-41-7	10,000
Bromine	7726-95-6	1,500
Chlorine	7782-50-5	1,500
Ethylene Oxide	75-21-8	5,000
Hydrogen Chloride	7647-01-0	5,000
Hydrogen Sulfide	7783-06-4	1,500
Isopropylamine	75-31-0	5,000
Ketene	463-51-4	100
Methylamine, Anhydrous	74-89-5	1,000
Methyl Chloride	74-87-3	15,000
Methyl Isocyanate	624-83-9	250
Nitric Acid ((94.5% by weight)	7697-37-2	500
Perchloromethyl Mercaptan	594-42-3	150
Perchloryl Fluoride	7616-94-6	5,000
Trifluorochloroethylene	79-38-9	10,000

Source: OSHA 1910.119, Appendix A

Qualified Investigator

The regulation clearly defines the investigating team:

> . . . at least one person knowledgeable in the process involved, including a contract employee if the incident involved work of the contractor, and other persons with appropriate knowledge and experience to thoroughly investigate and analyze the incident.

Time Requirements

OSHA regulations define specific time requirements for investigating any release or potential release of any chemical that is within the scope of 29 CFR 1910. The regulation states:

> An incident investigation shall be initiated as promptly as possible, but not later than 48 hours following the incident.

In part, the reason for this quick response to an incident is to assure that all pertinent evidence and facts can be preserved to facilitate the investigation.

Report Content

The regulation also defines specific topics that must be addressed in the report. These include

- Date of the incident,
- Date the investigation began,
- A complete description of the incident,
- Contributing factors, and
- Any recommendations resulting from the investigation.

Review Process

The regulation states:

> The report must be reviewed with all affected personnel whose job tasks are relevant to the incident findings, including contract employees where applicable.

The intent of this clause is to ensure that all affected employees understand why the incident occurred and what actions could prevent a recurrence. While the regulation does not specifically define the methods to comply with this requirement, it is imperative that the review be prompt and complete. In almost all cases, the review requires personal meetings, either individually or in small groups, to thoroughly review the incident and recommend corrective actions.

Corrective Actions

Regarding corrective actions, the regulation states:

> The employer shall establish a system to promptly address and resolve the incident-report findings and recommendations. Resolution and corrective actions shall be documented.

The regulation does not define *promptly* in definitive terms, but the intent is that all corrective actions must be implemented immediately.

The major difference between an OSHA-mandated investigation and other RCFA is that an appropriate corrective action or actions *must be implemented* as quickly as possible. In the non-OSHA-mandated RCFA process, a corrective action may or may not be implemented, depending on the results of the cost-benefit analysis.

The cost of corrective actions is not a consideration in the OSHA regulations, but it must be considered as part of the analysis. Because of the critical time line that governs an OSHA-mandated investigation, a full cost-benefit analysis may not be possible. However, *the investigating team should consider the cost-benefit impact of potential corrective actions.* The guidelines provided in the section on "Investigating a Reported Problem" in Chapter 3 should be followed as much as possible within the time constraints of the investigation.

OSHA's Investigation Process

Figure 3–16 illustrates the logic tree to follow for an OSHA-mandated investigation. While it is similar to other, nonmandated investigations, there are distinct differences. OSHA's Hazardous Waste Operations and Emergency Response (HAZWOPER) legislation protects workers who respond to emergencies, such as serious spills, involving hazardous materials. It also covers those employed in cleanup operations at uncontrolled hazardous waste sites and at EPA-licensed waste treatment, storage, and disposal facilities.

Emergency Planning and Community Right-to-Know Act

The Emergency Planning and Community Right-to-Know Act (EPCRA) is administered by the EPA and state and local agencies. It affects virtually all facilities that manufacture, use, or store hazardous chemicals. The following are the reporting requirements of the act:

- An inventory that includes the amount, nature, and location of any hazardous or extremely hazardous chemical present at a facility in an amount equal to or greater than its assigned "threshold level."
- Reports on releases of a "reportable quantity" of a listed hazardous substance, including the total annual releases during normal operations.

Resource Conservation and Recovery Act

The Resource Conservation and Recovery Act (RCRA) has a considerable number of regulations affecting spills in the workplace, including training of workers who might be expected to respond to them. RCRA is administered by the EPA.

U.S. Department of Transportation

Title 49 of the Code of Federal Regulations (CFR) addresses the U.S. DOT rules for packing and shipping of hazardous materials by air, road, rail, or water. CFR 49 covers issues such as appropriate containers, labeling, truck and railcar placards, and providing essential information that can aid in an emergency response in case of an incident involving hazardous materials.

Hazardous Materials Transportation Act

The Hazardous Materials Transportation Act (HMTA) defines *transportation releases* to be those that occur during loading, unloading, transportation or temporary storage of hazardous materials or waste. Releases that meet certain criteria (see Table 5–1) should be reported to the National Response Center (NRC) and the state response center. Most states also require calls to the local police or response agencies (often by calling 911).

Follow-up written reports are due within 30 days. If the carrier is required to report releases to the DOT, Form 5800.1 should be completed and sent to the following address:

> Information Systems Manager, DHM-63
> Research and Special Programs Administration
> U.S. Department of Transportation
> Washington, DC 20590-0001

CERCLA also requires certain spills to be reported by the owner or shipper to the NRC. To alert drivers and emergency responders to this requirement, the letters *RQ* must appear on shipping papers if the transporter is carrying, in one package, a substance on DOT's Hazardous Materials Table in an amount equal to or greater than the RQ shown in the table. Two additional requirements apply if hazardous waste is involved:

- Attach to the report a copy of the hazardous waste manifest.
- Include in the report an estimate of the quantity of the waste removed from the scene, the name and address of the facility to which it was taken, and the manner of disposition (Section IX of DOT Form F5800.1).

6

PROCESS PERFORMANCE

While the principles of RCFA can be used for almost any problem, the most prevalent application in an integrated plant is to resolve process-performance problems. These events normally do not result in catastrophic failure of critical plant systems or personnel injury, but they have a measurable, negative impact on the financial performance of the plant.

Resolution of deviations in process performance, such as operations problems (e.g., reduced product quality or capacity) or economics problems (e.g., high costs), often is more difficult than investigation of equipment failures or accidents.

Events like equipment failure or accidents have an absolute time of occurrence that facilitates an investigation. Once the event's time frame is established, the investigation can isolate all the variables and possible causes that may have contributed to the problem. Most process problems also have a unique timing associated with the deviation, but it is much harder to isolate. As a result, the investigation often is more difficult.

The recommended methodology for this type of RCFA is a cause-and-effect analysis. The technique of diagramming the potential causes of a specific event, such as product quality, loss of production capacity, or increase in operating costs, provides the structure and order needed to quickly and methodically resolve problems.

The cause-and-effect approach forces the investigator to identify each factor or variable that can contribute to a specific event. Once identified and graphically plotted in a fishbone diagram, the investigator or investigating team can quickly and easily evaluate each of the potential variables.

After the initial evaluation process, some of the initial factors will be eliminated. The investigating team should perform a cause-and-effect analysis on each of the remaining factors. In this evaluation, the factor being investigated becomes the *effect* in the fishbone diagram and the investigator must develop a comprehensive list of potential *causes* or factors that could cause that specific effect. This final step in the process should eliminate most of the potential causes of the initial problem or event being investigated. Those that remain should be the major contributors or root cause(s) of the problem.

OPERATIONS PROBLEMS

Common operations problems are product quality and capacity restrictions, and RCFA is an ideal way to resolve problems such as these.

Most product-quality problems are related either to equipment failure, changes in the operating envelope of plant systems, or poor practices. The logical approach provided by RCFA will resolve most, if not all, quality-related problems.

The initial clarification of a product-quality issue usually is the most difficult part of the investigation. Unlike equipment failure, where there is absolute evidence that something has happened, product-quality problems often are more abstract and less clearly defined. In some cases, quality problems may go undetected for days, weeks, or even months.

The second difficulty with quality-related RCFA is in isolating the specific point where the defects or deviations occurred. Unless a full-time quality inspection follows each step in the production process, quality defects normally are not detected until the production process is nearly complete. Because of this, it is difficult to quickly determine the specific plant area, process system, or machine that created the defect or deviation. As a result, the RCFA process must evaluate multiple process areas until the source of the problem can be absolutely isolated.

Capacity restrictions or loss of production capacity is another ideal application for RCFA. The logical, step-by-step approach used by this methodology, combined with its verification testing methods, provide a proven means to isolate the true cause of this type of problem.

Generally, capacity losses can be roughly isolated to a particular area of the plant. However, a thorough investigation may require evaluation of one or more of the production areas that precede the suspect process. In some cases, restriction in the prior processes may be the root cause of the perceived problem.

ECONOMIC PROBLEMS

RCFA methodology can be used to isolate and correct abnormal controllable costs within a plant. The basic approach defined in the preceding sections is valid, but may need to be adjusted to the specific application.

Part II

EQUIPMENT DESIGN EVALUATION GUIDE

This guide provides the basic design guidelines required to evaluate common plant machinery. Although it is not intended to be a comprehensive design-engineering manual, it provides the basic knowledge needed to understand the critical design issues that directly affect equipment reliability.

This document covers pumps, fans, blowers, fluidizers, conveyors, compressors, mixers and agitators, dust collectors, process rolls, gearboxes/ reducers, steam traps/condensate, inverters, control valves, and seals and packing.

7

PUMPS

Pumps are designed to transfer a specific volume of liquid at a particular pressure from a fixed source to a final destination in a process system. A pump's operating envelope is defined either by a hydraulic curve for centrifugal pumps or a pressure-volume (PV) diagram for positive-displacement pumps.

CENTRIFUGAL

Centrifugal pumps are highly susceptible to variations in process parameters, such as suction pressure, specific gravity of the pumped liquid, back pressure induced by control valves, and changes in demand volume. Therefore, the dominant reasons for centrifugal pump failures usually are process related.

Several factors dominate pump performance and reliability: internal configuration, suction condition, total dynamic pressure or head, hydraulic curve, brake horsepower, installation, and operating methods. These factors must be understood and used to evaluate any centrifugal pump-related problem or event.

Configuration

All centrifugal pumps are not configured alike. Variations in the internal configuration occur in the impeller type and orientation. These variations have a direct impact on a pump's stability, useful life, and performance characteristics.

Impeller Type

A variety of impeller types are used in centrifugal pumps. They range from simple radial flow, open designs to complex variable-pitch, high-volume, enclosed designs. Each type is designed to perform a specific function and should be selected with care.

In relatively small, general-purpose pumps, the impellers normally are designed to provide radial flow and the choices are limited to either an enclosed or open design.

Enclosed impellers are cast with the vanes fully encased between two disks. This type of impeller generally is used for clean, solid-free liquids. It has a much higher efficiency than the open design.

Open impellers have only one disk and the opposite side of the vanes is open to the liquid. Because of its lower efficiency, this design is limited to applications where slurries or solids are an integral part of the liquid.

Impeller Orientation

In single-stage centrifugal pumps, impeller orientation is fixed and not a factor in pump performance. However, it must be carefully considered in multistage pumps, which are available in two configurations: in-line and opposed. These configurations are illustrated in Figure 7–1.

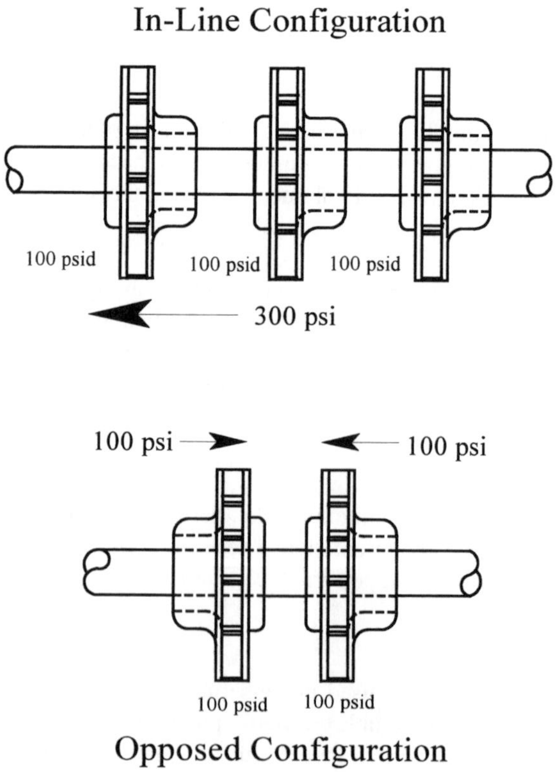

Figure 7–1 Impeller orientation of multistage centrifugal pumps.

In-Line   In-line configurations have all impellers facing in the same direction. As a result, the total differential pressure between the discharge and inlet is axially applied to the rotating element toward the outboard bearing. Because of this configuration, in-line pumps are highly susceptible to changes in the operating envelope.

Because of the tremendous axial pressures created by the in-line design, these pumps must have a positive means of limiting end play, or axial movement, of the rotating element. Normally, one of two methods is used to fix or limit axial movement: (1) a large thrust bearing is installed at the outboard end of the pump to restrict movement or (2) discharge pressure is vented to a piston mounted on the outboard end of the shaft.

Method 1 relies on the holding strength of the thrust bearing to absorb energy generated by the pump's differential pressure. If the process is reasonably stable, this design approach is valid and should provide a relatively trouble-free service life. However, this design cannot tolerate any radical or repeated variation in its operating envelope. Any change in the differential pressure or transient burst of energy generated by flow change will overload the thrust bearing, which may result in instantaneous failure.

Method 2 uses a bypass stream of pumped fluid at full discharge pressure to compensate for the axial load on the rotating element. While this design is more tolerant of process variations, it cannot compensate for repeated, instantaneous changes in demand, volume, or pressure.

Opposed   Multistage pumps that use opposed impellers are much more stable and can tolerate a broader range of process variables than those with an in-line configuration. In the opposed-impeller design, sets of impellers are mounted back to back on the shaft. As a result, the thrust or axial force generated by one of the pairs is canceled by the other. This design approach virtually eliminates axial forces. As a result, the pump requires no massive thrust bearing or balancing piston to fix the axial position of the shaft and rotating element.

Since the axial forces are balanced, this type of pump is much more tolerant of changes in flow and differential pressure than the in-line design. However, it is not immune to process instability or the transient forces caused by frequent radical changes in the operating envelope.

Performance

This section provides the basic knowledge needed to evaluate a centrifugal-pump application to determine its operating dynamics and identify any forcing function that may contribute to chronic reliability problems, premature failure, or loss of process performance.

Centrifugal pump performance is controlled primarily by two variables: suction conditions and total system pressure or head requirements. Total system pressure comprises the total vertical lift or elevation change, friction losses in the piping, and flow restrictions caused by the process. Other variables affecting performance include the pump's hydraulic curve and brake horsepower.

Suction Conditions

Factors affecting suction conditions are the net positive suction head (NPSH), suction volume, and entrained air or gas.

Net Positive Suction Head   Suction pressure, called the *net positive suction head or NPSH*, is a major factor governing pump performance. The variables affecting the suction head are shown in Figure 7–2.

Centrifugal pumps must have a minimum amount of consistent and constant positive pressure at the eye of its impeller. If this suction pressure is not available, the pump will be unable to transfer liquid. The suction supply can be open and below the pump's centerline, but the atmospheric pressure must be greater than the pressure required to lift the liquid to the impeller eye and provide the minimum NPSH required for proper pump operation.

At sea level, atmospheric pressure generates a pressure of 14.7 pounds per square inch (psi) to the surface of the supply liquid. This pressure minus vapor pressure, fric-

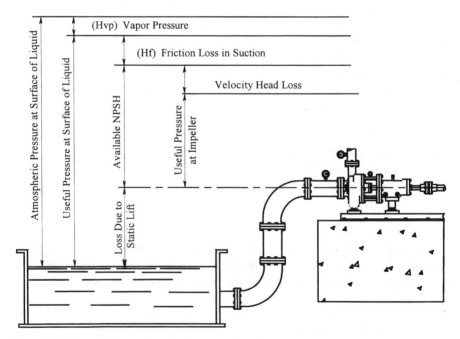

Figure 7–2 Net positive suction head in a suction lift application.

tion loss, velocity head, and static lift must be enough to provide the minimum NPSH requirements of the pump. The requirements vary with the volume of liquid transferred by the pump.

Most pump curves provide the minimum NPSH required for various flow conditions. This information, generally labeled NPSHR, usually is presented as a rising curve located near the bottom of the hydraulic curve. The data usually are expressed in *feet of head* rather than psi.

To convert from psi to feet of head for water, multiply by 2.31. For example, 14.7 psi is 14.7 × 2.31 or 33.957 feet of head. To convert feet of head to psi, multiply the total feet of head by 0.4331.

Suction Volume The pump's supply system must provide a consistent volume of single-phase liquid equal to or greater than the volume delivered by the pump. To accomplish this, the suction supply should have relatively constant volume and properties (e.g., pressure, temperature, specific gravity). Special attention must be paid in applications where the liquid has variable physical properties (e.g., specific gravity, density, viscosity). As the suction supply's properties vary, effective pump performance and reliability is adversely affected.

In applications where two or more pumps operate within the same system, special attention must be given to the suction flow requirements. Generally, these applications can be divided into two classifications: pumps in series and pumps in parallel.

Pumps in Series The suction conditions of two or more pumps in series are extremely critical (see Figure 7–3). Since each pump depends on the flow and pressure of the preceding pump, the flow characteristics must match. Both the flow and pressure must be matched to the required suction conditions of the next pump in the series.

For example, the first pump in the series may deliver 1,000 gpm and 100 ft of total dynamic head. The next pump in the series then will have an inlet volume of 1,000 gpm, but the inlet pressure will be 100 ft minus the pressure losses created by the total vertical lift between the two pumps' centerlines and all friction losses caused by the piping, valves, and the like.

This pressure at the suction of the second pump must be at least equal to its minimum NPSH operating requirements. If too low, the pump will cavitate and not generate sufficient volume and pressure for the process to operate properly.

Pumps in Parallel Pumps that operate in parallel normally share a common suction supply or discharge (or both). This is illustrated in Figure 7–4. Typically, a common manifold (i.e., pipe) or vessel is used to supply suction volume and pressure. The manifold's configuration must be such that all pumps receive adequate volume and net positive suction head. Special consideration must be given to flow patterns, friction losses, and restrictions.

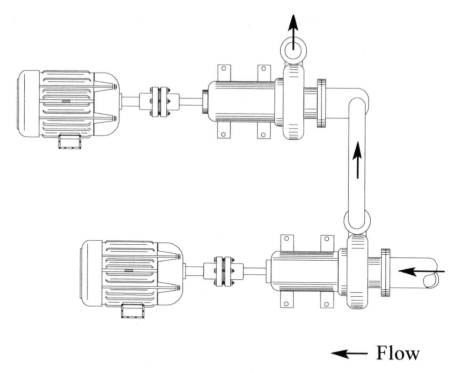

← Flow

Figure 7–3 Pumps in series must be properly matched.

One of the most common problems with pumps in parallel is suction starvation. This is caused by improper inlet piping, which permits more flow and pressure to reach one or more pumps but supplies insufficient quantities to the remaining pumps. In most cases, the condition results from poor piping or manifold design and may be expensive to correct.

Always remember that, when evaluating flow and pressure in pumping systems, they always will take the path of least resistance. For example, given a choice of flowing through a 6-in. pipe or a 2-in. pipe, most of the flow will go to the 6-in. pipe. Why? Simply because there is less resistance.

In parallel pump applications, there are two ways to balance the flow and pressure to the suction inlet of each pump. The first way is to design the piping so that the friction loss and flow path to each pump is equal. Although theoretically possible, this is extremely difficult to accomplish. The second method is to install a balancing valve in each suction line. By throttling or partially closing these valves, the system can be tuned to ensure proper flow and pressure to each pump.

Entrained Air or Gas Most pumps are designed to handle single-phase liquids within a limited range of specific gravities or viscosities. Entrainment of gases, such

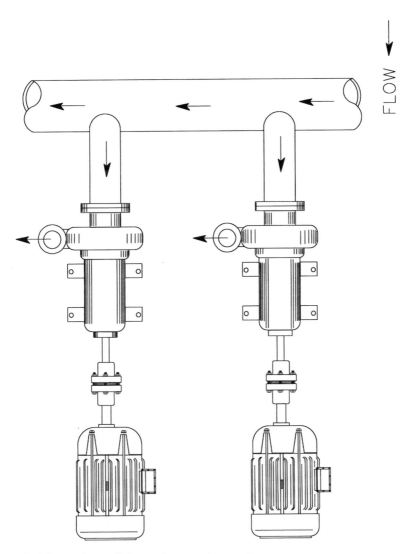

FLOW

Figure 7–4 Pumps in parallel may share suction supply.

as air or steam, has an adverse effect on both the pump's efficiency and its useful operating life. This is one form of cavitation, which is a common failure mode of centrifugal pumps. The typical causes of cavitation are leaks in suction piping and valves, or a change of phase induced by liquid temperature or suction pressure deviations. As an example, a one-pound suction pressure change in a boiler-feed application may permit the deaerator-supplied water to flash into steam. The introduction of a two-phase mixture of hot water and steam into the pump causes accelerated wear, instability, loss of pump performance, and chronic failure problems.

Total System Head

Centrifugal pump performance is controlled by the total system head (TSH) requirement, unlike positive-displacement pumps. TSH is defined as the total pressure
required to overcome all resistance at a given flow. This value includes all vertical lift,
friction loss, and back pressure generated by the entire system. It determines the efficiency, discharge volume, and stability of the pump.

Total Dynamic Head

The total dynamic head (TDH) is the difference between the discharge and suction
pressure of a centrifugal pump. This value is used by pump manufacturers to generate
hydraulic curves, such as those shown in Figures 7–5, 7–6, and 7–7. These curves rep-
resent the performance that can be expected for a particular pump under specific operating conditions. For example, a pump having a discharge pressure of 100 psig
(gauged pounds per square inch) and a positive pressure of 10 psig at the suction will
have a TDH of 90 psig.

Hydraulic Curve

Most pump hydraulic curves define pressure to be TDH rather than actual discharge
pressure. This is an important consideration when evaluating pump problems. For
example, a variation in suction pressure has a measurable impact on both the discharge pressure and the volume. Figure 7–5 is a simplified hydraulic curve for a single-stage, centrifugal pump. The vertical axis is TDH and the horizontal axis is the
discharge volume or flow.

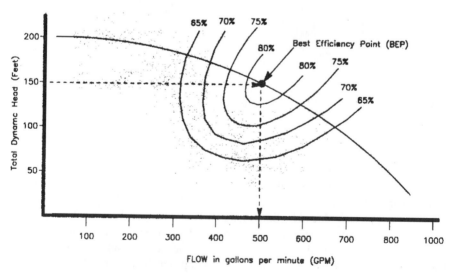

Figure 7–5 Simple hydraulic curve for centrifugal pump.

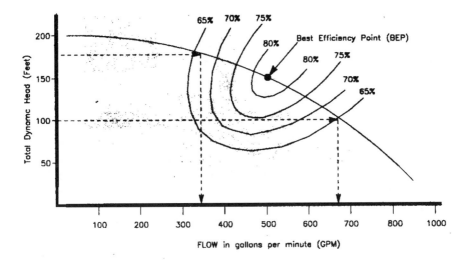

Figure 7–6 Actual centrifugal pump performance depends on total system head.

The best operating point for any centrifugal pump is called the *best efficiency point* (BEP). This is the point on the curve where the pump delivers the best combination of pressure and flow. In addition, the BEP specifies the point that provides the most stable pump operation with the lowest power consumption and longest maintenance-free service life.

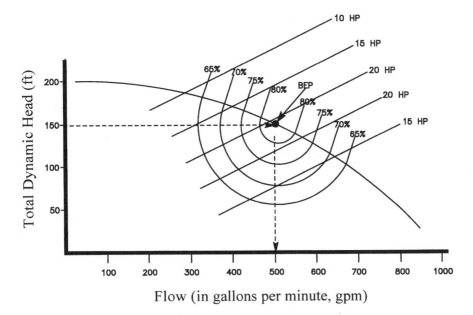

Figure 7–7 Brake horsepower needs change with process parameters.

In any installation, the pump will operate at the point where its TDH equals the TSH. When selecting a pump, it is hoped that the BEP is near the required flow where the TDH equals TSH on the curve. If not, there will be some operating-cost penalty as a result of the pump's inefficiency. This often is unavoidable because pump selection is determined by what is available commercially as opposed to selecting one that would provide the best theoretical performance.

For the centrifugal pump illustrated in Figure 7–5, the BEP occurs at a flow of 500 gpm with 150 ft TDH. If the TSH were increased to 175 ft, however, the pump's output would decrease to 350 gpm. Conversely, a decrease in TSH would increase the pump's output. For example, a TSH of 100 ft would result in a discharge flow of almost 670 gpm.

From an operating-dynamic standpoint, a centrifugal pump becomes more and more unstable as the hydraulic point moves away from the BEP. As a result, the normal service life decreases and the potential for premature failure of the pump or its components increases. A centrifugal pump should not be operated outside the efficiency range shown by the bands on its hydraulic curve, or 65 percent for the example shown in Figure 7–5.

If the pump is operated to the left of the minimum recommended efficiency point, it may not discharge enough liquid to dissipate the heat generated by the pumping operation. The heat that builds up within the pump can cause a catastrophic failure. This operating condition, called *shutoff*, is a leading cause of premature pump failure.

When the pump operates to the right of the last recommended efficiency point, it tends to overspeed and become extremely unstable. This operating condition, called *runout*, also can accelerate wear and bring on premature failure.

Brake Horsepower

Brake horsepower (BHP) refers to the amount of motor horsepower required for proper pump operation. The hydraulic curve for each type of centrifugal pump reflects its performance (i.e., flow and head) at various BHPs. Figure 7–7 is an example of a simplified hydraulic curve that includes the BHP parameter.

Note the diagonal lines that indicate the BHP required for various process conditions. For example, the pump illustrated in Figure 7–7 requires 22.3 horsepower at its BEP. If the TSH required by the application increases from 150 ft to 175 ft, the horsepower required by the pump will increase to 24.6. Conversely, when the TSH decreases, the required horsepower also decreases. The brake horsepower required by a centrifugal pump can be easily calculated by

$$\text{Brake Horsepower} = \frac{\text{Flow (gpm)} \times \text{Specific Gravity} \times \text{Total Dynamic Head (ft)}}{3960 \times \text{Efficiency}}$$

With two exceptions, the certified hydraulic curve for any centrifugal pump provides the data required to calculate the actual brake horsepower. Those exceptions are specific gravity and TDH.

Specific gravity must be determined for the particular liquid being pumped. For example, water has a specific gravity of 1.0. Most other clear liquids have a specific gravity of less than 1.0. Slurries and other liquids that contain solids or are highly viscous materials generally have a higher specific gravity. Reference books, like Ingersoll Rand's *Cameron Hydraulic Databook*, provide these values for many liquids.

The TDH can be measured directly for any application using two calibrated pressure gauges. Install one gauge in the suction inlet of the pump and the another on the discharge. The difference between these two readings is the TDH.

With the actual TDH, flow can be determined directly from the hydraulic curve. Simply locate the measured pressure on the hydraulic curve by drawing a horizontal line from the vertical axis (i.e., TDH) to a point where it intersects the curve. From the intersection point, draw a vertical line downward to the horizontal axis (i.e., flow). This provides an accurate flow rate for the pump.

The intersection point also provides the pump's efficiency for that specific point. Since the intersection may not fall exactly on one of the efficiency curves, some approximation may be required.

Installation

Centrifugal pump installation should follow the Hydraulic Institute standards, which provide specific guidelines to prevent distortion of the pump and its baseplate. Distortions can result in premature wear, loss of performance, or catastrophic failure. The following should be evaluated as part of a root cause failure analysis: foundation, piping support, and inlet and discharge piping configurations.

Foundation

Centrifugal pumps require a rigid foundation that prevents torsional or linear movement of the pump and its baseplate. In most cases, this type of pump is mounted on a concrete pad having enough mass to securely support the baseplate, which has a series of mounting holes. Depending on size, there may be three to six mounting points on each side.

The baseplate must be securely bolted to the concrete foundation at all these points. One common installation error is to leave out the center baseplate lag bolts. This permits the baseplate to flex with the torsional load generated by the pump.

Piping Support

Pipe strain causes the pump casing to deform and results in premature wear or failure. Therefore, both suction and discharge piping must be adequately supported to prevent strain. In addition, flexible isolator connectors should be used on both suction and discharge pipes to ensure proper operation.

Inlet-Piping Configuration

Centrifugal pumps are highly susceptible to turbulent flow. The Hydraulic Institute provides guidelines for piping configurations that are specifically designed to ensure laminar flow of the liquid as it enters the pump. As a general rule, the suction pipe should provide a straight, unrestricted run that is six times the inlet diameter of the pump.

Installations that have sharp turns, shutoff or flow-control valves, or undersized pipe on the suction-side of the pump are prone to chronic performance problems. Such deviations from good engineering practices result in turbulent suction flow and cause hydraulic instability that severely restricts pump performance.

Discharge-Piping Configuration

The restrictions on discharge piping are not as critical as for suction piping, but using good engineering practices ensures longer life and trouble-free operation of the pump. The primary considerations that govern discharge-piping design are friction losses and total vertical lift or elevation change. The combination of these two factors is called TSH, discussed in the section earlier in this chapter, which represents the total force that the pump must overcome to perform properly. If the system is designed properly, the TDH of the pump will equal the TSH at the desired flow rate.

In most applications, it is relatively straightforward to confirm the total elevation change of the pumped liquid. Measure all vertical rises and drops in the discharge piping, then calculate the total difference between the pump's centerline and the final delivery point.

Determining the total friction loss, however, is not as simple. Friction loss is caused by a number of factors, and all depend on the flow velocity generated by the pump. The major sources of friction loss include

- Friction between the pumped liquid and the sidewalls of the pipe.
- Valves, elbows, and other mechanical flow restrictions.
- Other flow restrictions, such as back pressure created by the weight of liquid in the delivery storage tank or resistance within the system component that uses the pumped liquid.

A number of reference books, like Ingersoll-Rand's *Cameron Hydraulics Databook*, provide the pipe-friction losses for common pipes under various flow conditions. Generally, data tables define the approximate losses in terms of specific pipe lengths or runs. Friction loss can be approximated by measuring the total run length of each pipe size used in the discharge system, dividing the total by the equivalent length used in the table, and multiplying the result by the friction loss given in the table.

Each time the flow is interrupted by a change of direction, a restriction caused by valving, or a change in pipe diameter, the flow resistance of the piping increases substantially. The actual amount of this increase depends on the nature of the restriction. For example, a short-radius elbow creates much more resistance than a long-radius elbow, a ball valve's resistance is much greater than a gate valve's, and the resistance from a pipe-size reduction of 4 in. will be greater than for a 1-in. reduction. Reference tables are available in hydraulics handbooks that provide the relative values for each of the major sources of friction loss. As in the friction tables mentioned previously, these tables often provide the friction loss as equivalent runs of straight pipe.

In some cases, friction losses are difficult to quantify. If the pumped liquid is delivered to an intermediate storage tank, the configuration of the tank's inlet determines if it adds to the system pressure. If the inlet is on or near the top, the tank will add no back pressure. However, if the inlet is below the normal liquid level, the total height of liquid above the inlet must be added to the total system head.

In applications where the liquid is used directly by one or more system components, the contribution of these components to the total system head may be difficult to calculate. In some cases, the vendor's manual or the original design documentation will provide this information. If these data are not available, then the friction losses and back pressure need to be measured or an overcapacity pump selected for service based on a conservative estimate.

Operating Methods

Normally, little consideration is given to operating practices for centrifugal pumps. However, some critical practices must be followed, such as using proper startup procedures, using proper bypass operations, and operating under stable conditions.

Startup Procedures

Centrifugal pumps always should be started with the discharge valve closed. As soon as the pump is activated, the valve should be opened slowly to its full-open position.

The only exception to this rule is when there is positive back pressure on the pump at startup. Without adequate back pressure, the pump will absorb a substantial torsional load during the initial startup sequence. The normal tendency is to overspeed because there is no resistance on the impeller.

Bypass Operation

Many pump applications include a bypass loop intended to prevent deadheading (i.e., pumping against a closed discharge). Most bypass loops consist of a metered orifice inserted in the bypass piping to permit a minimal flow of liquid. In many cases, the flow permitted by these metered orifices is not sufficient to dissipate the heat generated by the pump or to permit stable pump operation.

If a bypass loop is used, it must provide sufficient flow to assure reliable pump operation. The bypass should provide sufficient volume to permit the pump to operate within its designed operating envelope. This envelope is bound by the efficiency curves that are included on the pump's hydraulic curve, which provides the minimum flow required to meet this requirement.

Stable Operating Conditions

Centrifugal pumps cannot absorb constant, rapid changes in operating environment. For example, frequent cycling between full-flow and no-flow assures premature failure of any centrifugal pump. The radical surge of back pressure generated by rapidly closing a discharge valve, referred to as *hydraulic hammer*, generates an instantaneous shock load that actually can tear the pump from its piping and foundation.

In applications where frequent changes in flow demand are required, the pump system must be protected from such transients. Two methods can be used to protect the system:

- Slow down the transient. Instead of instant valve closing, throttle the system over a longer time interval. This will reduce the potential for hydraulic hammer and prolong pump life.
- Install proportioning valves. For applications where frequent radical flow swings are necessary, the best protection is to install a pair of proportioning valves that have inverse logic. The primary valve controls flow to the process. The second controls flow to a full-flow bypass. Because of their inverse logic, the second valve will open in direct proportion as the primary valve closes, keeping the flow from the pump nearly constant.

Positive Displacement

Centrifugal and positive-displacement pumps share some basic design requirements. Both require an adequate, constant suction volume to deliver designed fluid volumes and liquid pressures to their installed systems. In addition, both are affected by variations in the liquid's physical properties (e.g., specific gravity, viscosity) and flow characteristics through the pump.

Unlike centrifugal pumps, positive-displacement pumps are designed to displace a specific volume of liquid each time they complete one cycle of operation. As a result, they are less prone to variations in performance as a direct result of changes in the downstream system. However, there are exceptions to this. Some types of positive-displacement pumps, such as screw-types, are extremely sensitive to variations in system back pressure. Causes of this sensitivity were discussed previously in this chapter.

When positive-displacement pumps are used, the system must be protected from excessive pressures. This type of pump will deliver whatever discharge pressure is required to overcome the system's total head. The only restrictions on its maximum pressure are the burst pressure of the system's components and the maximum driver horsepower.

As a result of their ability to generate almost unlimited pressure, all positive-displacement pumps' systems must be fitted with relief valves on the downstream side of the discharge valve. This is required to protect the pump and its discharge piping from overpressurization. Some designs include a relief valve integral to the pump's housing. Others use a separate valve installed in the discharge piping.

Positive-displacement pumps deliver a definite volume of liquid for each cycle of pump operation. Therefore, the only factor, except for pipe blockage, that affects the flow rate in an ideal positive-displacement pump is the speed at which it operates. The flow resistance of the system in which the pump is operating does not affect the flow rate through the pump. Figure 7–8 shows the characteristics curve (i.e., flow rate versus head) for a positive-displacement pump.

The dashed line in Figure 7–8 shows the actual positive-displacement pump performance. This line reflects the fact that, as the discharge pressure of the pump increases, liquid leaks from the discharge back to the suction-inlet side of the pump casing. This reduces the pump's effective flow rate. The rate at which liquid leaks from the pump's discharge to its suction side is called *slip*. Slip is the result of two primary factors: (1) design clearance required to prevent metal-to-metal contact of moving parts and (2) internal part wear.

Minimum design clearance is necessary for proper operation, but it should be enough to minimize wear. Proper operation and maintenance of positive-displacement pumps limits the amount of slip caused by wear.

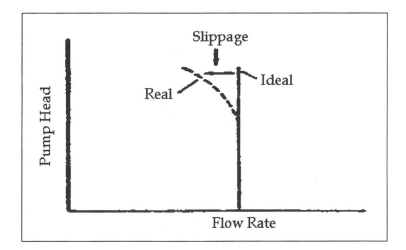

Figure 7–8 Positive-displacement pump characteristics curve (Mobley 1989).

Configuration

Positive-displacement pumps come in a variety of configurations. Each has a specific function and should be selected based on the effectiveness and reliability in a specific application. The major types of positive-displacement pumps are gear, screw, vane, and lobe.

Gear

The most common type of positive-displacement pump uses a combination of gears and configurations to provide the liquid pressure and volume required by the application. Variations of gear pumps are spur, helical, and herringbone.

Spur   The simple spur-gear pump shown in Figure 7–9 consists of two spur gears meshing together and revolving in opposite directions within a casing. Only a few thousandths-of-an-inch clearance exists between the case, gear faces, and teeth extremities. This design forces any liquid filling the space bounded by two successive gear teeth and the case to move with the teeth as they revolve. When the gear teeth mesh with the teeth of the other gear, the space between them is reduced. This forces the entrapped liquid out through the pump's discharge pipe.

As the gears revolve and the teeth disengage, the space again opens on the suction side of the pump, trapping new quantities of liquid and carrying it around the pump case to the discharge. Lower pressure results as the liquid moves away from the suction side, which draws liquid in through the suction line.

For gears having a large number of teeth, the discharge is relatively smooth and continuous, with small quantities of liquid delivered to the discharge line in rapid succes-

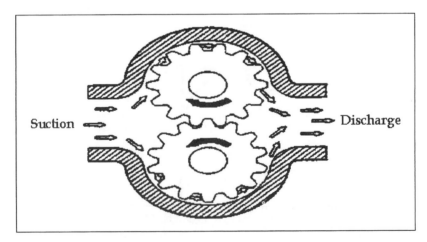

Figure 7–9 Simple spur gear pump (Mobley 1989).

sion. For gears having fewer teeth, the space between them is greater and the capacity increases for a given speed. However, this increases the tendency to have a pulsating discharge.

In all simple-gear pumps, power is applied to one of the gear shafts, which transmits power to the driven gear through their meshing teeth. No valves are in the gear pump to cause friction losses as in the reciprocating pump. The high impeller velocities required in centrifugal pumps, which result in friction losses, are not needed in gear pumps. This makes gear pumps well suited for viscous fluids, such as fuel and lubricating oils.

Helical The helical-gear pump is a modification of the spur-gear pump and has certain advantages. With a spur gear, the entire length of the tooth engages at the same time. With a helical gear, the point of engagement moves along the length of the tooth as the gear rotates. This results in a steadier discharge pressure and less pulsation than in a spur-gear pump.

Herringbone The herringbone-gear pump is another modification of the simple-gear pump. The principal difference in operation from the simple-gear pump is that the pointed center section of the space between two teeth begins discharging fluid before the divergent outer ends of the preceding space complete discharging. This overlapping tends to provide a steadier discharge pressure. The power transmission from the driving gear to the driven gear also is smoother and quieter.

Screw

There are many design variations for screw-type, positive-displacement rotary pumps. The primary variations are the number of intermeshing screws, the screw pitch, and fluid-flow direction.

The most common type of screw pump consists of two screws mounted on two parallel shafts that mesh with close clearances. One screw has a right-handed thread, while the other has a left-handed. One shaft drives the other through a set of timing gears, which synchronize the screws and maintain clearance between them.

The screws rotate in closely fitting duplex cylinders that have overlapping bores. While all clearances are small, no contact occurs between the two screws or between the screws and the cylinder walls. The complete assembly and the usual flow path for such a pump are shown in Figure 7–10.

In this type of pump, liquid is trapped at the outer end of each pair of screws. As the first space between the screw threads rotates away from the opposite screw, a spiral-shaped quantity of liquid is enclosed when the end of the screw again meshes with the opposite screw. As the screw continues to rotate, the entrapped spiral of liquid slides along the cylinder toward the center discharge space while the next slug is entrapped. Each screw functions similarly, and each pair of screws discharges an equal quantity

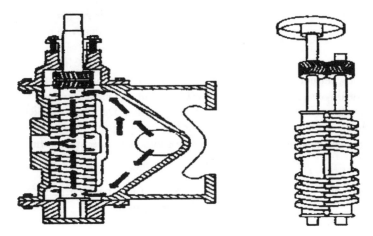

Figure 7–10 Two-screw, low-pitch screw pump.

of liquid in opposed streams toward the center, thus eliminating hydraulic thrust. The removal of liquid from the suction end by the screws produces a reduction in pressure, which draws liquid through the suction line.

Vane

The sliding-vane pump shown in Figure 7–11, another type of positive-displacement pump, is used with viscous fluids. It consists of a cylindrical bored housing with a

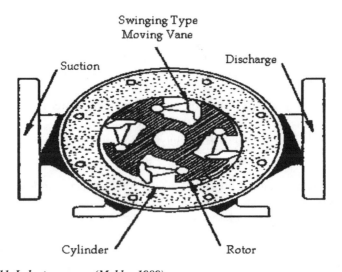

Figure 7–11 Lobe-type pump (Mobley 1989).

suction inlet on one side and a discharge outlet on the other. A cylindrical-shaped rotor having a diameter smaller than the cylinder is driven about an axis position above the cylinder's centerline. The clearance between the rotor and the top of the cylinder is small, but it increases toward the bottom.

The rotor has vanes that move in and out as it rotates, maintaining a sealed space between the rotor and the cylinder wall. The vanes trap liquid on the suction side and carry it to the discharge side, where contraction of the space expels it through the discharge line. The vanes may swing on pivots or slide in slots in the rotor.

Lobe

The lobe-type pump shown in Figure 7–12 is another variation of the simple-gear pump. It can be considered a simple-gear pump having only two or three lobes per rotor. Other than this difference, its operation and the function of its parts are no different. Some designs of lobe pumps are fitted with replaceable gibs, or thin plates, carried in grooves at the extremity of each lobe where they make contact with the casing. Gibs promote tightness and absorb radial wear.

Performance

Positive-displacement pump performance is determined by three primary factors: liquid viscosity, rotating speed, and suction supply.

Viscosity

Positive-displacement pumps are designed to handle viscous liquids such as oil, grease, and polymers. However, a change in viscosity has a direct effect on its performance. As

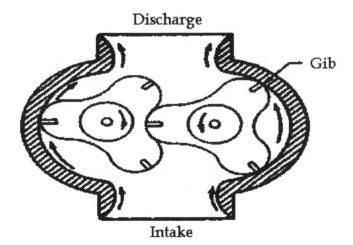

Figure 7–12 Rotary sliding-vane pump (Mobley 1989).

the viscosity increases, the pump must work harder to deliver a constant volume of fluid to the discharge. As a result, the brake horsepower needed to drive the pump increases to keep the rotating speed constant and prevent a marked reduction in the volume of liquid delivered to the discharge. If the viscosity change is great enough, the brake horsepower requirements may exceed the capabilities of the motor.

Temperature variation is the major contributor to viscosity change. The design specifications should define an acceptable range of both viscosity and temperature for each application. These two variables are closely linked and should be clearly understood.

Rotating Speed

With positive-displacement pumps, output is directly proportional to the rotating speed. If the speed changes from its normal design point, the volume of liquid delivered also will change.

Suction Supply

To a degree, positive-displacement pumps are self-priming. In other words, they have the ability to draw liquid into their suction ports. However, they must have a constant volume of liquid available. Therefore, the suction-supply system should be designed to ensure that a constant volume of nonturbulent liquid is available to each pump in the system.

Pump performance and its useful operating life are enhanced if the suction-supply system provides a consistent positive pressure. Pumps required to overcome suction lift must work harder to deliver product to the discharge.

Installation

Installation requirements for positive-displacement pumps are basically the same as those for centrifugal pumps. Those requirements were discussed previously in this chapter.

Special attention should be given to the suction-piping configuration. Poor piping practices in hydraulic-system applications are primary sources of positive-displacement pump problems, particularly in parallel pump applications. Often the suction piping does not provide adequate volume to each pump in parallel configurations.

Operating Methods

If a positive-displacement pump is properly installed, there are few restrictions on operating methods. The primary operating concerns are bypass operation and speed-change rates.

8

FANS, BLOWERS, AND FLUIDIZERS

Technically, fans and blowers are separate types of devices that have a similar function. However, the terms often are used interchangeably to mean any device that delivers a quantity of air or gas at a desired pressure. Differences between these two devices are their rotating elements and discharge-pressure capabilities. Fluidizers are identical to single-stage, screw-type compressors or blowers.

CENTRIFUGAL FANS

Centrifugal fans are one of the most common machines used in industry. They utilize a rotating element with blades, vanes, or propellers to extract or deliver a specific volume of air or gas. The rotating element is mounted on a rotating shaft that must provide the energy required to overcome inertia, friction, and other factors that restrict or resist air or gas flow in the application. These generally are low-pressure machines designed to overcome friction and either suction or discharge-system pressure.

Configuration

Centrifugal fans can be classified by the type of rotating element or wheel that is used to move the air or gas. The major classifications are propeller and axial. Axial fans can be further differentiated by blade configuration.

Propeller

This type of fan consists of a propeller, or paddle-wheel, mounted on a rotating shaft within a ring, panel, or cage. The most widely used propeller fans perform light- or medium-duty functions, such as in ventilation units where air can be moved in any direction. These fans commonly are used in wall mountings to inject air into or exhaust air from a space. Figure 8–1 illustrates a belt-driven propeller fan appropriate for medium-duty applications.

This type of fan has a limited ability to boost pressure. Its use should be limited to applications where the total resistance to flow is less than 1 in. of water. In addition, it should not be used in corrosive environments or where explosive gases are present.

Axial

Axial fans are essentially propeller fans enclosed within a cylindrical housing or shroud. They can be mounted inside ductwork or a vessel housing to inject or exhaust air or gas.

These fans have an internal motor mounted on spokes or struts to centralize the unit within the housing. Electrical connections and grease fittings are mounted externally on the housing. Arrow indicators on the housing show the direction of air flow and

Figure 8–1 Belt-driven propeller fan for medium duty applications (Higgins and Mobley 1995).

rotation of the shaft, which enables the unit to be correctly installed in the duct work. Figure 8–2 shows the inlet end of a direct-connected, tube-axial fan.

This type of fan should not be used in corrosive or explosive environments since the motor and bearings cannot be protected. Applications where concentrations of airborne abrasives are present should be avoided as well.

Axial fans use three primary types of blades or vanes: backward-curved, forward-curved, and radial. Each type has specific advantages and disadvantages.

Backward-Curved Blades The backward-curved blade provides the highest efficiency and lowest sound level of all axial-type, centrifugal fan blades. Its advantages include

- Moderate to high volume.
- Static pressure range up to approximately 30 in. of water (gauge).
- Highest efficiency of any type of fan.
- Lowest noise level of any fan for the same pressure and volumetric requirements.
- Self-limiting BHP characteristics. (Motors can be selected to prevent overload at any volume and the BHP curve rises to a peak and then declines as volume increases.)

Figure 8–2 Inlet end of a direct-connected tube-axial fan (Higgins and Mobley 1995).

The limitations of backward-curved blades are

- Weighs more and occupies considerably more space than other designs of equal volume and pressure.
- Large wheel width.
- Not to be used in dusty environments or where sticky or stringy materials are used because residues adhering to the blade surface cause imbalance and eventual bearing failure.

Forward-Curved Blades This design is commonly referred to as a *squirrel-cage fan*. The unit has a wheel with a large number of wide, shallow blades; a very large intake area relative to the wheel diameter; and a relatively slow operational speed. The advantages of forward-curved blades include

- Excellent for any volume at low to moderate static pressure using clean air.
- Occupies approximately same space as backward-curved blade fan.
- More efficient and much quieter during operation than propeller fans for static pressures above approximately 1 in. of water (gauge).

The limitations of forward-curved blades include

- Not as efficient as backward-curved blade fans.
- Should not be used in dusty environments or handle sticky or stringy materials that could adhere to the blade surface.
- BHP increases as this fan approaches maximum volume as opposed to backward-curved blade centrifugal fans, which experience a decrease in BHP as they approach maximum volume.

Radial Blades Industrial exhaust fans fall into this category. The design is rugged and may be belt driven or driven directly by a motor. The blade shape varies considerably from flat to various bent configurations to increase efficiency slightly or to suit particular applications. The advantages of radial-blade fans include

- Best suited for severe duty, especially when fitted with flat radial blades.
- Simple construction that lends itself to easy field maintenance.
- Highly versatile industrial fan that can be used in extremely dusty environments as well as clean air.
- Appropriate for high-temperature service.
- Handles corrosive or abrasive materials.

The limitations of radial-blade fans include

- Lowest efficiency in centrifugal-fan group.
- Highest sound level in centrifugal-fan group.
- BHP increases as fan approaches maximum volume.

Performance

A fan is inherently a constant-volume machine. It operates at the same volumetric flow rate (i.e., cubic feet per minute, or cfm) when operating in a fixed system at a constant speed, regardless of changes in air density. However, the pressure developed and the horsepower required vary directly with the air density.

The following factors affect centrifugal-fan performance: brake horsepower, fan capacity, fan rating, outlet velocity, static efficiency, static pressure, tip speed, mechanical efficiency, total pressure, velocity pressure, natural frequency, and suction (inlet) conditions. Some of these factors are used in the mathematical relationships referred to as *fan laws*, which are discussed later in the chapter.

Brake Horsepower

Brake horsepower (BHP) is the power input required by the fan shaft to produce the required volumetric flow rate (cfm) and pressure.

Fan Capacity

The fan capacity (FC) is the volume of air moved per minute by the fan (cfm). NOTE: The density of air is 0.075 pounds per cubic foot at atmospheric pressure and 68°F.

Fan Rating

The fan rating predicts the fan's performance at one operating condition, which includes the fan size, speed, capacity, pressure, and horsepower.

Outlet Velocity

The outlet velocity (OV, feet per minute) is the number of cubic feet of gas moved by the fan per minute divided by the inside area of the fan outlet, or discharge area, in square feet.

Static Efficiency

Static efficiency (SE) is not the true mechanical efficiency, but is convenient to use in comparing fans. This is calculated by the following equation:

$$\text{Static Efficiency (SE)} = \frac{0.000157 \times \text{FC} \times \text{SP}}{\text{BHP}}$$

Static Pressure

Static pressure (SP) generated by the fan can exist whether the air is in motion or trapped in a confined space. SP is always expressed in inches of water (gauge).

Tip Speed

The tip speed (TS) is the peripheral speed of the fan wheel in feet per minute (fpm).

$$\text{Tip Speed} = \text{Rotor Diameter} \times \pi \times \text{Rotations per Minute (rpm)}$$

Mechanical Efficiency

True mechanical efficiency (ME) is equal to the total input power divided by the total output power.

Total Pressure

Total pressure (TP), in inches of water (gauge), is the sum of the velocity pressure and static pressure.

Velocity Pressure

Velocity pressure (VP) is produced by the fan only when the air is moving. Air having a velocity of 4,000 fpm exerts a pressure of 1 in. of water (gauge) on a stationary object in its flow path.

Natural Frequency

General-purpose fans are designed to operate below their first natural frequency. In most cases, the fan vendor will design the rotor-support system so that the rotating element's first critical speed is between 10 and 15 percent above the rated running speed. While this practice is questionable, it is acceptable if the design speed and rotating-element mass are maintained. However, if either of these two factors changes, there is a high probability that serious damage or premature failure will result.

Inlet-air Conditions

As with centrifugal pumps, fans require stable inlet (suction) conditions. Ductwork should be configured to ensure an adequate volume of clean air or gas, stable inlet pressure, and laminar flow. If the supply air is extracted from the environment, it is subject to variations in moisture, dirt content, barometric pressure, and density. However, these variables should be controlled as much as possible. As a minimum, inlet filters should be installed to minimize the amount of dirt and moisture that enters the fan.

Excessive moisture and particulates have an extremely negative impact on fan performance and cause two major problems: abrasion or tip wear and plate out. High concentrations of particulate matter in the inlet air act as abrasives that accelerate fan-rotor wear. In most cases, however, this wear is restricted to the high-velocity areas of the rotor, such as the vane or blade tips, but it can affect the entire assembly.

Plate out is a much more serious problem. The combination of particulates and moisture can form a "glue" that binds to the rotor assembly. As this contamination builds up on the rotor, the assembly's mass increases, which reduces its natural frequency. If enough plate out occurs, the fan's rotational speed may coincide with the rotor's reduced natural frequency. With a strong energy source like the running speed, excitation of the rotor's natural frequency can result in catastrophic fan failure. Even if catastrophic failure does not occur, the vibration energy generated by the fan may cause bearing damage.

Fan Laws

The mathematical relationships referred to as *fan laws* can be useful when applied to fans operating in a fixed system or to geometrically similar fans. However, caution should be exercised when using these relationships. They apply only to identical fans and applications. The basic laws are

- Volume in cubic feet per minute (cfm) varies directly with the rotating speed (rpm).
- Static pressure varies with the rotating speed squared (rpm^2).
- Brake horsepower (BHP) varies with the speed cubed (rpm^3).

The fan-performance curves shown in Figures 8–3 and 8–4 show the performance of fans of the same type but designed for different volumetric flow rates, operating in the same duct system handling air at the same density.

Curve #1 is for a fan designed to handle 10,000 cfm in a duct system whose calculated system resistance is determined to be 1 in. water (gauge). This fan will operate at the point where the fan pressure (SP) curve intersects the system resistance curve (TSH). This intersection point is called the *point of rating*. The fan will operate at this

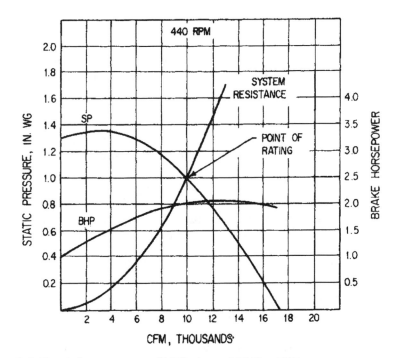

Figure 8–3 Fan performance curve #1 (Higgins and Mobley 1995).

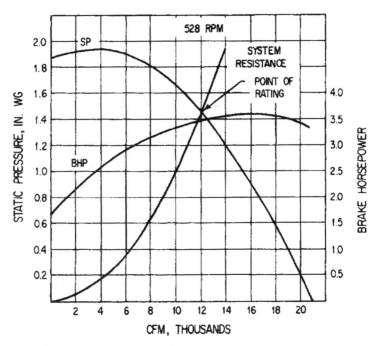

Figure 8–4 Fan performance curve #2 (Higgins and Mobley 1995).

point provided the fan's speed remains constant and the system's resistance does not change. The system resistance curve illustrates that the resistance varies as the square of the volumetric flow rate (cfm). The BHP of the fan required for this application is 2 hp.

Curve #2 illustrates the situation if the fan's design capacity is increased by 20 percent, increasing output from 10,000 to 12,000 cfm. Applying the fan laws, the calculations are

$$
\begin{aligned}
\text{New rpm} &= 1.2 \times 440 \\
&= 528 \text{ rpm (20\% increase)} \\
\text{New SP} &= 1.2 \times 1.2 \times 1 \text{ in. water (gauge)} \\
&= 1.44 \text{ in. (44\% increase)} \\
\text{New TSH} &= \text{New SP} = 1.44 \text{ in.} \\
\text{New BHP} &= 1.2 \times 1.2 \times 1.2 \times 2 \\
&= 1.73 \times 2 \\
&= 3.46 \text{ hp (73\% increase)}
\end{aligned}
$$

The curve representing the system resistance is the same in both cases, since the system has not changed. The fan will operate at the same relative point of rating and will move the increased volume through the system. The mechanical and static efficiencies are unchanged.

The increased brake horsepower (BHP) required to drive the fan is a very important point to note. If the Curve #1 fan had been driven by a 2-hp motor, the Curve #2 fan needs a 3.5-hp motor to meet its volumetric requirement.

Centrifugal-fan selection is based on rating values such as air flow, rpm, air density, and cost. Table 8–1 is a typical rating table for a centrifugal fan. Table 8–2 provides air-density ratios.

Installation

Proper fan installation is critical to reliable operation. Suitable foundations, adequate bearing-support structures, properly sized ductwork, and flow-control devices are the primary considerations.

Foundations

As with any other rotating machine, fans require a rigid, stable foundation. With the exception of in-line fans, they must have a concrete footing or pad that is properly sized to provide a stable footprint and prevent flexing of the rotor-support system.

Bearing-Support Structures

In most cases, with the exception of in-line configurations, fans are supplied with a vendor-fabricated base. Bases normally consist of fabricated metal stands that support the motor and fan housing. The problem with most of the fabricated bases is that they lack the rigidity and stiffness to prevent flexing or distortion of the fan's rotating element. The typical support structure is composed of a relatively light-gauge material ($^3/_{16}$ in.) and lacks the cross-bracing or structural stiffeners needed to prevent distortion of the rotor assembly. Because of this limitation, many plants fill the support structure with concrete or another solid material.

However, this approach does little to correct the problem. When the concrete solidifies, it pulls away from the sides of the support structure. Without direct bonding and full contact with the walls of the support structure, stiffness is not significantly improved.

The best solution to this problem is to add cross-braces and structural stiffeners. If they are properly sized and affixed to the support structure, the stiffness can be improved and rotor distortion reduced.

Ductwork

Ductwork should be sized to provide minimal friction loss throughout the system. Bends, junctions with other ductwork, and any change of direction should provide a

Table 8–1 Typical Rating Table for a Centrifugal Fan

CFM	OV	1/4" SP		3/8" SP		1/2" SP		5/8" SP		3/4" SP	
		RPM	BHP	RPM	BHP	RPM	BHP	RPM	BHP	RPM	BHP
7458	800	262	0.45	289	0.60	314	0.75	337	0.92	360	1.09
8388	900	281	0.55	305	0.72	330	0.89	351	1.06	372	1.25
9320	1000	199	0.68	325	0.85	347	1.04	368	1.23	387	1.43
10252	1100	319	0.79	343	1.00	365	1.21	385	1.42	403	1.63
11184	1200	338	0.93	362	1.17	383	1.40	402	1.63	420	1.85
12116	1300	358	1.10	381	1.35	402	1.61	421	1.85	438	2.10
13048	1400	379	1.29	401	1.55	421	1.83	439	2.10	456	2.37
13980	1500	401	1.50	420	1.78	440	2.08	458	2.37	475	2.66
14912	1600	422	1.74	441	2.03	459	2.35	477	2.57	494	2.98
15844	1700	444	2.01	462	2.32	479	2.65	496	2.98	513	3.32
18776	1800	467	2.31	483	2.63	499	2.97	516	3.33	532	3.68
17708	1900	489	2.65	504	2.98	520	3.33	536	3.70	551	4.07
18840	2000	512	3.02	526	3.36	541	3.72	556	4.10	571	4.49
19572	2100	535	3.43	548	3.77	562	4.15	576	4.53	590	4.95
20504	2200	558	3.87	570	4.23	584	4.61	597	5.02	610	5.43
21436	2300	582	4.36	593	4.72	605	5.12	618	5.54	631	5.95
22368	2400	505	4.89	616	5.26	627	5.67	640	6.10	652	6.54
23300	2500	628	5.45	639	5.85	650	6.26	651	6.70	673	7.16
24232	2600	652	6.09	662	6.48	672	6.90	683	7.34	694	7.81
25164	2700	676	6.75	685	7.15	695	7.58	705	8.04	715	8.52
26096	2800	700	7.47	708	7.88	718	8.32	727	8.78	738	9.27
27028	2900	723	8.24	732	8.66	741	9.11	750	9.58	760	10.08

Table 8–1 Typical Rating Table for a Centrifugal Fan (continued)

1/4" SP		3/8" SP		1/2" SP		5/8" SP		3/4" SP	
RPM	BHP	RPM	BHP	RPM	BHP	RPM	BHP	RPM	BHP
382	1.27	403	1.45	444	1.85	483	2.28	520	2.73
393	1.44	413	1.63	451	2.05	488	2.49	523	2.96
406	1.63	425	1.83	461	2.27	495	2.72	529	3.21
421	1.84	439	2.06	473	2.51	505	2.99	537	3.50
438	2.08	454	2.31	486	2.79	517	3.29	547	3.81
455	2.34	471	2.59	501	3.09	531	3.62	559	4.16
473	2.63	489	2.90	517	3.43	545	3.98	572	4.54
491	2.94	506	3.23	534	3.79	551	4.37	587	4.96
509	3.28	524	3.58	552	4.19	578	4.79	603	5.41
528	3.64	542	3.97	570	4.61	595	5.25	619	5.90
547	4.03	561	4.38	588	5.06	613	5.74	637	6.42
566	4.45	580	4.81	606	5.54	631	6.28	654	6.98
585	4.89	599	5.28	625	6.05	649	6.81	672	7.57
604	5.36	618	5.78	644	6.59	668	7.40	690	8.19
624	5.87	637	6.30	663	7.16	686	8.01	708	8.85
644	6.41	657	6.86	682	7.77	705	8.66	727	9.54
664	6.99	677	7.46	701	8.41	724	9.36	746	10.27
685	7.63	697	8.10	721	9.09	743	10.07	765	11.04
706	8.30	717	8.77	740	9.80	762	10.83	784	11.84
727	9.01	738	9.53	760	10.55	782	11.63	803	12.69
748	9.78	759	10.30	780	11.35	801	12.48	822	13.57
770	10.60	780	11.13	800	12.20	821	13.35	841	14.49

Source: Unknown

Table 8–2 Air Density Ratios

Air Temp °F	Altitude, feet above sea level												
	0	1,000	2,000	3,000	4,000	5,000	6,000	7,000	8,000	9,000	10,000	15,000	20,000
	Barometric pressure, inches of mercury												
	29.92	28.86	27.82	26.82	25.84	24.90	23.98	23.09	22.22	21.39	20.58	16.89	13.75
70	1.000	0.964	0.930	0.896	0.864	0.832	0.801	0.772	0.743	0.714	0.688	0.564	0.460
100	0.946	0.912	0.880	0.848	0.818	0.787	0.758	0.730	0.703	0.676	0.651	0.534	0.435
150	0.869	0.838	0.808	0.770	0.751	0.723	0.696	0.671	0.646	0.620	0.598	0.490	0.400
200	0.803	0.774	0.747	0.720	0.694	0.668	0.643	0.620	0.596	0.573	0.552	0.453	0.369
250	0.747	0.720	0.694	0.669	0.645	0.622	0.598	0.576	0.555	0.533	0.514	0.421	0.344
300	0.697	0.672	0.648	0.624	0.604	0.580	0.558	0.538	0.518	0.498	0.480	0.393	0.321
350	0.654	0.631	0.608	0.586	0.565	0.544	0.524	0.505	0.486	0.467	0.450	0.369	0.301
400	0.616	0.594	0.573	0.552	0.532	0.513	0.493	0.476	0.458	0.440	0.424	0.347	0.283
450	0.582	0.561	0.542	0.522	0.503	0.484	0.466	0.449	0.433	0.416	0.401	0.328	0.268
500	0.552	0.532	0.513	0.495	0.477	0.459	0.442	0.426	0.410	0.394	0.380	0.311	0.254
550	0.525	0.506	0.488	0.470	0.454	0.437	0.421	0.405	0.390	0.375	0.361	0.296	0.242
600	0.500	0.482	0.465	0.448	0.432	0.416	0.400	0.386	0.372	0.352	0.344	0.282	0.230
650	0.477	0.460	0.444	0.427	0.412	0.397	0.382	0.368	0.354	0.341	0.328	0.269	0.219
700	0.457	0.441	0.425	0.410	0.395	0.380	0.366	0.353	0.340	0.326	0.315	0.258	0.210

Source: Unknown

clean, direct flow path. All ductwork should be airtight and leak free to ensure proper operation.

Flow-Control Devices

Fans should always have inlet and outlet dampers or other flow-control devices, such as variable-inlet vanes. Without them, it is extremely difficult to match fan performance to actual application demand. The reason for this difficulty is that a number of variables (e.g., usage, humidity, and temperature) directly affect the input-output demands for each fan application. Flow-control devices provide the means to adjust fan operation for actual conditions. Figure 8–5 shows an outlet damper with stream-

Figure 8–5 Outlet damper with streamlined blades and linkage arranged to move adjacent blades in opposite directions for even throttling (Higgins and Mobley 1995).

lined blades and linkage arranged to move adjacent blades in opposite directions for even throttling.

Air-flow controllers must be inspected frequently to ensure that they are fully operable and operate in unison with each other. They also must close tightly. Ensure that the control indicators show the precise position of the vanes in all operational conditions. The "open" and "closed" positions should be permanently marked and visible at all times. Periodic lubrication of linkages is required.

Turn-buckle screws on the linkages for adjusting flow rates should never be moved without first measuring the distance between the set-point markers on each screw. This is important if the adjustments do not produce the desired effect and you wish to return to the original settings.

Operating Methods

Because fans are designed for stable, steady-state operation, variations in speed or load may have an adverse effect on their operating dynamics. The primary operating method that should be understood is output control. Two methods can be used to control fan output: dampers and fan speed.

Dampers

Dampers can be used to control the output of centrifugal fans within the effective control limits. Centrifugal fans have a finite acceptable control range, typically about 15 percent above and below its design point. Control variations outside this range severely affect the reliability and useful life of the fan.

The recommended practice is to use an inlet damper rather than a discharge damper for this control function whenever possible. Restricting the inlet with suction dampers can effectively control the fan's output. When using dampers to control fan performance, however, caution should be exercised to ensure that any changes remain within the fan's effective control range.

Fan Speed

Varying fan speed is an effective means of controlling a fan's performance. As defined by the fan laws (discussed earlier), both volume and pressure can be controlled directly by changing the rotating speed of the fan. However, caution must be used when changing fan speed. All rotating elements, including fan rotors, have one or more critical speeds. When the fan's speed coincides with one of the critical speeds, the rotor assembly becomes extremely unstable and could fail catastrophically.

In most general purpose applications, fans are designed to operate between 10 and 15 percent below their first critical speed. If speed is increased on these fans, there is a good potential for a critical-speed problem. Other applications have fans designed to operate between their first and second critical speeds. In this instance, any change up or down may cause the speed to coincide with one of the critical speeds.

BLOWERS

A blower uses mating helical lobes or screws and is utilized for the same purpose as a fan. These are normally moderate- to high-pressure devices. Blowers are almost identical both physically and functionally to positive-displacement compressors. Therefore, compressor information found in Chapter 4 on basic design criteria, physical laws, and operating characteristics applies to blowers as well.

FLUIDIZERS

Fluidizers are identical to single-stage, screw-type compressors or blowers. They are designed to provide moderate- to high-pressure transfer of nonabrasive, dry materials.

9

CONVEYORS

Conveyors are used to transport materials from one location to another within a plant or facility. The variety of conveyor systems is almost infinite, but the two major classifications used in typical chemical plants are pneumatic and mechanical. Note that the power requirements of a pneumatic-conveyor system are much greater than for a mechanical conveyor of equal capacity. However, both systems offer some advantages.

PNEUMATIC

Pneumatic conveyors are used to transport dry, free-flowing, granular material in suspension within a pipe or duct. This is accomplished by the use of a high-velocity air-stream or by the energy of expanding compressed air within a comparatively dense column of fluidized or aerated material. Principal uses are (1) dust collection; (2) conveying soft materials, such as flake or tow; and (3) conveying hard materials, such as fly ash, cement, and sawdust.

The primary advantages of pneumatic-conveyor systems are the flexibility of piping configurations and their ability to greatly reduce the explosion hazard. Pneumatic conveyors can be installed in almost any configuration required to meet the specific application. With the exception of the primary driver, there are no moving parts that can fail or cause injury. However, when used to transport explosive materials, the potential for static charge buildup that could cause an explosion remains.

Configuration

A typical pneumatic-conveyor system consists of Schedule-40 pipe or ductwork, which provides the primary flow path used to transport the conveyed material. Motive power is provided by the primary driver, which can be either a fan, fluidizer, or positive-displacement compressor.

Performance

Pneumatic conveyor performance is determined by the following factors: primary-driver output, internal surface of the piping or ductwork, and the condition of the transported material. Specific factors affecting performance include motive power, friction loss, and flow restrictions.

Motive Power

The motive power is provided by the primary driver, which generates the gas (typically air) velocity required to transport material within a pneumatic-conveyor system. Therefore, the efficiency of the conveying system depends on the primary driver's operating condition.

Friction Loss

Friction loss within a pneumatic-conveyor system is a primary source of efficiency loss. The piping or ductwork must be properly sized to minimize friction without lowering the velocity below the value needed to transport the material.

Flow Restrictions

An inherent weakness of pneumatic-conveyor systems is their potential for blockage. The inside surfaces must be clean and free of protrusions or other defects that can restrict or interrupt the flow of material. In addition, when a system is shut down or the velocity drops below the minimum required to keep the transported material suspended, the product will drop out or settle in the piping or ductwork. In most cases, this settled material will compress and lodge in the piping. The restriction caused by this compacted material will reduce flow and eventually result in a complete blockage of the system.

Another major contributor to flow restrictions is blockage caused by system backups. This occurs when the end point of the conveyor system (i.e., storage silo, machine, or vessel) cannot accept the entire delivered flow of material. As the transported material backs up in the conveyor piping, it compresses and forms a solid plug that must be removed manually.

Installation

All piping and ductwork should be as straight and short as possible. Bends should have a radius of at least three diameters of the pipe or ductwork. The diameter should be selected to minimize friction loss and maintain enough velocity to prevent settling of the conveyed material. Branch lines should be configured to match as closely as possible the primary flow direction and avoid 90° angles to the main line. The area of the main conveyor line at any point along its run should be 20 to 25 percent greater than the sum of all its branch lines.

When vertical runs are short in proportion to the horizontal runs, the size of the riser can be restricted to provide the additional velocity, if needed. If the vertical runs are long, the primary or a secondary driver must provide sufficient velocity to transport the material.

Cleanouts, or drop-legs, should be installed at regular intervals throughout the system to permit foreign materials to drop out of the conveyed material. In addition, they provide the means to remove materials that drop out when the system is shut down or air velocity is lost. It is especially important to install adequate cleanout systems near flow restrictions and at the end of the conveyor system.

Operating Methods

Pneumatic-conveyor systems must be operated properly to prevent chronic problems, with the primary concern being to maintain constant flow and velocity. If either of these variables is permitted to drop below the system's design envelope, partial or complete blockage of the conveyor system will occur.

Constant velocity can be maintained only when the system is operated within its performance envelope and when regular cleanout is part of the normal operating practice. In addition, the primary driver must be in good operating condition. Any deviation in the primary driver's efficiency reduces the velocity and can result in partial or complete blockage.

The entire pneumatic-conveyor system should be completely evacuated before shutdown to prevent material from settling in the piping or ductwork. In noncontinuous applications, the conveyor system should be operated until all material within the conveyor's piping is transported to its final destination. Material that is allowed to settle will compact and partially block the piping. Over time, this will cause a total blockage of the conveyor system.

MECHANICAL

A variety of mechanical-conveyor systems are used in chemical plants. These systems generally incorporate chain- or screw-type mechanisms.

Chain

A commonly used chain-type system is a flight conveyor (e.g., Hefler conveyor), which is used to transport granular, lumpy, or pulverized materials along a horizontal or inclined path within totally enclosed ductwork. The Hefler systems generally have lower power requirements than the pneumatic conveyor and, in addition, prevent product contamination. This section focuses primarily on the Hefler-type conveyor because it is one of the most commonly used systems.

Table 9–1 Approximate Capacities of Hefler Conveyors

Flight Width and Depth (in.)	Quantity of Material (ft³/ft)	Approximate Capacity (short tons/hour)	Lump Size, Single Strand (in.)	Lump Size, Dual Strand (in.)
12 × 6	0.40	60	31.5	4.0
15 × 6	0.49	73	41.5	5.0
18 × 6	0.56	84	5.0	6.0
24 × 8	1.16	174		10.0
30 × 10	1.60	240		14.0
36 × 12	2.40	360		16.0

Source: Theodore Baumeister, ed., *Marks' Standard Handbook for Mechanical Engineers*, 8th ed. (New York: McGraw-Hill, 1978).

Configuration

The Hefler-type conveyor uses a center- or double-chain configuration to provide positive transfer of material within its ductwork. Both chain configurations use hardened bars or U-shaped devices that are an integral part of the chain to drag the conveyed material through the ductwork.

Performance

Data used to determine Hefler conveyors' capacity and the size of material that can be conveyed are presented in Table 9–1. Note that the data are for level conveyors. When conveyors are inclined, the capacity data obtained from Table 9-1 must be multiplied by the factors provided in Table 9–2.

Installation

The primary installation concerns with Hefler-type conveyor systems are the ductwork and primary-drive system.

Ductwork The inside surfaces of the ductwork must be free of defects or protrusions that interfere with the movement of the conveyor's chain or transported product. This is especially true at the joints. The ductwork must be sized to provide adequate chain

Table 9–2 Capacity Correction Factors for Inclined Hefler Conveyors

Inclination, degrees	20	25	30	35
Factor	0.9	0.8	0.7	0.6

Source: Theodore Baumeister, ed., *Marks' Standard Handbook for Mechanical Engineers*, 8th ed. (New York: McGraw-Hill, 1978).

clearance but should not be large enough to have areas where the chain-drive bypasses the product.

A long horizontal run followed by an upturn is inadvisable because of radial thrust. All bends should have a large radius to permit smooth transition and prevent material buildup. As with pneumatic conveyors, the ductwork should include cleanout ports at regular intervals for ease of maintenance.

Primary Drive System   Most mechanical conveyors use a primary-drive system that consists of an electric motor and a speed-increaser gearbox. See Chapter 14 for more information on gear-drive performance and operation criteria.

The drive-system configuration may vary, depending on the specific application or vendor. However, all configurations should include a single point-of-failure device, such as a shear pin, to protect the conveyor. The shear pin is critical in this type of conveyor because it is prone to catastrophic failure caused by blockage or obstructions that may lock the chain. Use of the proper shear pin prevents major damage to the conveyor system.

For continuous applications, the primary-drive system must have adequate horsepower to handle a fully loaded conveyor. Horsepower requirements should be determined based on the specific product's density and the conveyor's maximum-capacity rating.

For intermittent applications, the initial startup torque is substantially greater than for continuous operation. Therefore, selection of the drive system and the designed failure point of the shear device must be based on the maximum startup torque of a fully loaded system.

If either the drive system or designed failure point is not properly sized, this type of conveyor is prone to chronic failure. The predominant types of failure are frequent breakage of the shear device and trips of the motor's circuit breaker caused by excessive startup amp loads.

Operating Methods

Most mechanical conveyors are designed for continuous operation and may exhibit problems in intermittent-service applications. The primary problem is the startup torque for a fully loaded conveyor. This is especially true for conveyor systems handling material that tends to compact or compress on settling in a vessel, such as the conveyor trough.

The only positive method of preventing excessive startup torque is to ensure that the conveyor is completely empty before shutdown. In most cases, this can be accomplished by isolating the conveyor from its supply for a few minutes prior to shutdown. This time delay permits the conveyor to deliver its entire load of product before it is shut off.

In applications where it is impossible to completely evacuate the conveyor prior to shutdown, the only viable option is to jog, or step start, the conveyor. Step starting reduces the amp load on the motor and should control the torque to prevent the shear pin from failing.

If, instead of step starting, the operator applies full motor load to a stationary, fully loaded conveyor, one of two things will occur: (1) the drive motor's circuit breaker will trip as a result of excessive amp load or (2) the shear pin installed to protect the conveyor will fail. Either of these failures adversely affects production.

Screw

The screw, or spiral, conveyor is widely used for pulverized or granular, noncorrosive, nonabrasive materials in systems requiring moderate capacities, distances no more than about 200 feet, and moderate inclines (≤35°). It usually costs substantially less than any other type of conveyor and can be made dust tight by installing a simple cover plate.

Abrasive or corrosive materials can be handled with suitable construction of the helix and trough. Conveyors using special materials, such as hard-faced cast iron and linings or coatings, on the components that come into contact with the materials can be specified in these applications. The screw conveyor will handle lumpy material if the lumps are not large in proportion to the diameter of the screw's helix.

Screw conveyors may be inclined. A standard-pitch helix will handle material on inclines up to 35°. Capacity is reduced in inclined applications, and Table 9–3 provides the approximate reduction in capacity for various inclines.

Configuration

Screw conveyors have a variety of configurations. Each is designed for specific applications or materials. Standard conveyors have a galvanized-steel rotor, or helix, and trough. For abrasive and corrosive materials (e.g., wet ash), both the helix and trough may be hard-faced cast iron. For abrasives, the outer edge of the helix may be faced with a renewable strip of Stellite(tm) (a cobalt alloy produced by Haynes Stellite Co.) or other similarly hard material. Aluminum, bronze, Monel, or stainless steel also may be used to construct the rotor and trough.

Table 9–3 Screw Conveyor Capacity Reductions for Inclined Applications

Inclination, degrees	10	15	20	25	30	35
Reduction in capacity, %	10	26	45	58	70	78

Source: Theodore Baumeister, ed., *Marks' Standard Handbook for Mechanical Engineers*, 8th ed. (New York: McGraw-Hill, 1978).

Short-Pitch Screw

The standard helix used for screw conveyors has a pitch approximately equal to its outside diameter. The short-pitch screw is designed for applications with inclines greater than 29°.

Variable-Pitch Screw

Variable-pitch screws having the short pitch at the feed-end automatically control the flow to the conveyor and correctly proportion the load down the screw's length. Screws having what is referred to as a *short section*, which has either a shorter pitch or smaller diameter, are self-loading and do not require a feeder.

Cut Flight

Cut-flight conveyors are used for conveying and mixing cereals, grains, and other light material. They are similar to normal flight or screw conveyors, and the only difference is the configuration of the paddles or screw. Notches are cut in the flights to improve the mixing and conveying efficiency when handling light, dry materials.

Ribbon Screw

Ribbon screws are used for wet and sticky materials, such as molasses, hot tar, and asphalt. This type of screw prevents the materials from building up and altering the natural frequency of the screw. A buildup can cause resonance problems and possibly catastrophic failure of the unit.

Paddle Screw

The paddle-screw conveyor is used primarily for mixing materials like mortar and paving mixtures. An example of a typical application is churning ashes and water to eliminate dust.

Performance

Process parameters, such as density, viscosity, and temperature, must be constantly maintained within the conveyor's design operating envelope. Slight variations can affect performance and reliability. In intermittent applications, extreme care should be taken to fully evacuate the conveyor prior to shutdown. In addition, caution must be exercised when restarting a conveyor in case an improper shutdown was performed and material was allowed to settle.

Power Requirements

The horsepower requirement for the conveyor-head shaft, H, for horizontal screw conveyors can be determined from the following equation:

$$H = (ALN + CWLF) \times 10 - 6$$

Table 9–4 Factor A for Self-Lubricating Bronze Bearings

Conveyor Diameter, in.	6	9	10	12	14	16	18	20	24
Factor A		54	96	114	171	255	336	414	510	690

Source: Theodore Baumeister, ed., *Marks' Standard Handbook for Mechanical Engineers*, 8th ed. (New York: McGraw-Hill, 1978).

where

A = factor for size of conveyor (see Table 9–4);
C = material volume, ft^3/h;
F = material factor, unitless (see Table 9–5);
L = length of conveyor, ft;
N = conveyor rotation speed (rpm);
W = density of material, lb/ft^3.

Table 9–5 Power Requirements by Material Group

Material Group	Max. Cross-Section (%) Occupied by the Material	Max. Density of Material (lb/ft^3)	Max. rpm for 6-in. diameter	Max. rpm for 20-in. diameter
1	45	50	170	110
2	38	50	120	75
3	31	75	90	60
4	25	100	70	50
5	12 1/2		30	25

Group 1: F factor is 0.5 for light materials such as barley, beans, brewers, grains (dry), coal (pulv.), corn meal, cottonseed meal, flaxseed, flour, malt, oats, rice, wheat.

Group 2: Includes fines and granular material. The values of F are alum (pulv.), 0.6; coal (slack or fines), 0.9; coffee beans, 0.4; sawdust, 0.7; soda ash (light), 0.7; soybeans, 0.5; fly ash, 0.4.

Group 3: Includes materials with small lumps mixed with fines. Values of F are alum, 1.4; ashes (dry), 4.0; borax, 0.7; brewers grains (wet), 0.6; cottonseed, 0.9; salt, coarse or fine, 1.2; soda ash (heavy), 0.7.

Group 4: Includes semiabrasive materials, fines, granular and small lumps. Values of F are acid phosphate (dry), 1.4; bauxite (dry), 1.8; cement (dry), 1.4; clay, 2.0; fuller's earth, 2.0; lead salts, 1.0; lime-stone screenings, 2.0; sugar (raw), 1.0; white lead, 1.0; sulfur (lumpy), 0.8; zinc oxide, 1.0.

Group 5: Includes abrasive lumpy materials which must be kept from contact with hanger bearings. Values of F are wet ashes, 5.0; flue dirt, 4.0; quartz (pulv.), 2.5; silica sand, 2.0; sewage sludge (wet and sandy), 6.0.

Source: Theodore Baumeister, ed., *Marks' Standard Handbook for Mechanical Engineers*, 8th ed. (New York: McGraw-Hill, 1978).

Table 9–6 Allowance Factor

H (hp)	1	1–2	2–4	4–5	5
G	2	1.5	1.25	1.1	1.0

Source: Theodore Baumeister, ed., *Marks' Standard Handbook for Mechanical Engineers*, 8th ed. (New York: McGraw-Hill, 1978).

In addition to H, the motor size depends on the drive efficiency (E) and a unitless allowance factor (G), which is a function of H. Values for G are found in Table 9–6. The value for E usually is 90 percent.

$$\text{Motor hp} = HG/E$$

Table 9–5 gives the information needed to estimate the power requirement: percentages of helix loading for five groups of material, maximum material density or capacity, allowable speeds for 6-in. and 20-in. diameter screws, and the factor F.

Volumetric Efficiency

Screw-conveyor performance also is determined by the volumetric efficiency of the system. This efficiency is determined by the amount of slip or bypass generated by the conveyor. The amount of slip in a screw conveyor is determined primarily by three factors: product properties, screw efficiency, and clearance between the screw and the conveyor barrel or housing.

Product Properties   Not all materials or products have the same flow characteristics. Some have plastic characteristics and flow easily. Others do not self-adhere and tend to separate when pumped or conveyed mechanically. As a result, the volumetric efficiency is directly affected by the properties of each product. This also affects screw performance.

Screw Efficiency   Each of the common screw configurations (i.e., short pitch, variable pitch, cut flights, ribbon, and paddle) has varying volumetric efficiencies, depending on the type of product conveyed. Screw designs or configurations must be carefully matched to the product to be handled by the system.

For most medium- to high-density products in a chemical plant, the variable-pitch design normally provides the highest volumetric efficiency and lowest required horsepower. Cut-flight conveyors are highly efficient for light, nonadhering products, such as cereals, but are inefficient when handling heavy, cohesive products. Ribbon conveyors are used to convey heavy liquids, such as molasses, but are not very efficient and have a high slip ratio.

Clearance Improper clearance is the source of many volumetric-efficiency problems. It is important to maintain proper clearance between the outer ring, or diameter, of the screw and the conveyor's barrel, or housing, throughout the operating life of the conveyor. Periodic adjustments to compensate for wear, variations in product, and changes in temperature are essential. While the recommended clearance varies with specific conveyor design and the product to be conveyed, excessive clearance has a severe impact on conveyor performance as well.

Installation

Installation requirements vary greatly with screw-conveyor design. The vendor's operating and maintenance (O&M) manuals should be consulted and followed to ensure proper installation. However, as with practically all mechanical equipment, some basic installation requirements are common to all screw conveyors. Installation requirements presented here should be evaluated in conjunction with the vendor's O&M manual. If the information provided here conflicts with the vendor-supplied information, the O&M manual's recommendations always should be followed.

Foundation 0

The conveyor and its support structure must be installed on a rigid foundation that absorbs the torsional energy generated by the rotating screws. Because of the total overall length of most screw conveyors, a single foundation that supports the entire length and width should be used. There must be enough lateral (i.e., width) stiffness to prevent flexing during normal operation. Mounting conveyor systems on decking or suspended-concrete flooring should provide adequate support.

Support Structure Most screw conveyors are mounted above the foundation level on a support structure that generally has a slight downward slope from the feed end to the discharge end. While this improves the operating efficiency of the conveyor, it also may cause premature wear of the conveyor and its components.

The support's structural members (i.e., I-beams and channels) must be adequately rigid to prevent conveyor flexing or distortion during normal operation. Design, sizing, and installation of the support structure must guarantee rigid support over the full operating range of the conveyor. When evaluating the structural requirements, variations in product type, density, and operating temperature also must be considered. Since these variables directly affect the torsional energy generated by the conveyor, the worst-case scenario should be used to design the conveyor's support structure.

Product-Feed System A major limiting factor of screw conveyors is their ability to provide a continuous supply of incoming product. While some conveyor designs, such as those having a variable-pitch screw, provide the ability to self-feed, their installation should include a means of ensuring a constant, consistent incoming supply of product.

In addition, the product-feed system must prevent entrainment of contaminates in the incoming product. Normally, this requires an enclosure that seals the product from outside contaminants.

Operating Methods

As previously discussed, screw conveyors are sensitive to variations in incoming product properties and the operating environment. Therefore, the primary operating concern is to maintain a uniform operating envelope at all times, in particular by controlling variations in incoming product and operating environment.

Incoming-Product Variations   Any measurable change in the properties of the incoming product directly affects the performance of a screw conveyor. Therefore, the operating practices should limit variations in product density, temperature, and viscosity. If they occur, the conveyor's speed should be adjusted to compensate for them.

For property changes directly related to product temperature, preheaters or coolers can be used in the incoming-feed hopper and heating or cooling traces can be used on the conveyor's barrel. These systems provide a means of achieving optimum conveyor performance despite variations in incoming product.

Operating-Environment Variations   Changes in the ambient conditions surrounding the conveyor system may also cause deviations in performance. A controlled environment will substantially improve the conveyor's efficiency and overall performance. Therefore, operating practices should include ways to adjust conveyor speed and output to compensate for variations. The conveyor should be protected from wind chill, radical variations in temperature and humidity, and any other environment-related variables.

10

COMPRESSORS

A compressor is a machine used to increase the pressure of a gas or vapor. Compressors can be grouped into two major classifications: centrifugal and positive displacement. This chapter provides a general discussion of these types of compressors.

CENTRIFUGAL

In general, the centrifugal designation is used when the gas flow is radial and the energy transfer is predominantly due to a change in the centrifugal forces acting on the gas. The force utilized by the centrifugal compressor is the same as that utilized by centrifugal pumps.

In a centrifugal compressor, air or gas at atmospheric pressure enters the eye of the impeller. As the impeller rotates, the gas is accelerated by the rotating element within the confined space created by the volute of the compressor's casing. The gas is compressed as more gas is forced into the volute by the impeller blades. The pressure of the gas increases as it is pushed through the reduced free space within the volute.

As in centrifugal pumps, there may be several stages to a centrifugal air compressor. In these multistage units, a progressively higher pressure is produced by each stage of compression.

Configuration

The actual dynamics of centrifugal compressors are determined by their design. Common designs are overhung or cantilever, centerline, and bullgear.

Overhung or Cantilever

The cantilever design is more susceptible to process instability than centerline centrifugal compressors. Figure 10–1 illustrates a typical cantilever design.

The overhung design of the rotor (i.e., no outboard bearing) increases the potential for radical shaft deflection. Any variation in laminar flow, volume, or load of the inlet or discharge gas forces the shaft to bend or deflect from its true centerline. As a result, the mode shape of the shaft must be monitored closely.

Centerline

Centerline designs (i.e., horizontal and vertical split case) are more stable over a wider operating range but should not be operated in a variable-demand system. Figure 10–2 illustrates the normal airflow pattern through a horizontal split-case compressor. Inlet air enters the first stage of the compressor, where pressure and velocity increases occur. The partially compressed air is routed to the second stage, where the velocity and pressure are increased further. This process can be continued by adding additional stages until the desired final discharge pressure is achieved.

Two factors are critical to the operation of these compressors: impeller configuration and laminar flow, which must be maintained through all of the stages.

Figure 10–1 Cantilever centrifugal compressor is susceptible to instability (Gibbs 1971).

The impeller configuration has a major impact on stability and the operating envelope. There are two impeller configurations: in-line and back-to-back, or opposed. With the in-line design, all impellers face in the same direction. With the opposed design, impeller direction is reversed in adjacent stages.

In-Line A compressor with all impellers facing in the same direction generates substantial axial forces. The axial pressures generated by each impeller for all the stages are additive. As a result, massive axial loads are transmitted to the fixed bearing. Because of this load, most of these compressors use either a Kingsbury thrust bearing or a balancing piston to resist axial thrusting. Figure 10–3 illustrates a typical balancing piston.

All compressors that use in-line impellers must be monitored closely for axial thrusting. If the compressor is subjected to frequent or constant unloading, the axial clearance will increase due to this thrusting cycle. Ultimately, this frequent thrust loading will lead to catastrophic failure of the compressor.

Opposed By reversing the direction of alternating impellers, the axial forces generated by each impeller or stage can be minimized. In effect, the opposed impellers tend to cancel the axial forces generated by the preceding stage. This design is more stable and should not generate measurable axial thrusting, which allows these units to contain a normal float and fixed rolling-element bearing.

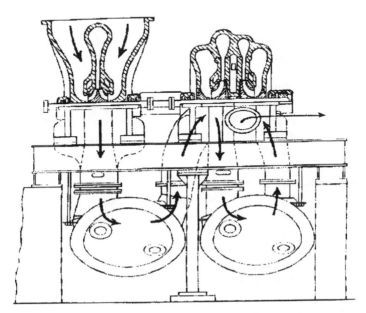

Figure 10–2 Airflow through a centerline centrifugal compressor (Gibbs 1971).

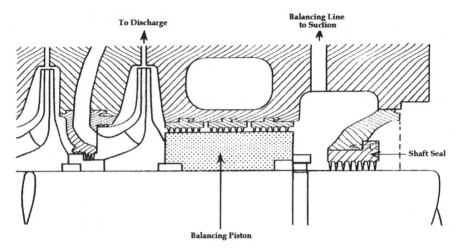

Figure 10–3 Balancing piston resists axial thrust from the in-line impeller design of a centerline centrifugal compressor (Gibbs 1971).

Bullgear

The bullgear design uses a direct-driven helical gear to transmit power from the primary driver to a series of pinion-gear-driven impellers located around the circumference of the bullgear. Figure 10–4 illustrates a typical bullgear compressor layout.

The pinion shafts typically have a cantilever-type design that has an enclosed impeller on one end and a tilting-pad bearing on the other. The pinion gear is between these two components. The number of impeller-pinions (i.e., stages) varies with the application and the original equipment vendor. However, all bullgear compressors contain multiple pinions that operate in series.

Atmospheric air or gas enters the first-stage pinion, where the pressure is increased by the centrifugal force created by the first-stage impeller. The partially compressed air leaves the first stage, passes through an intercooler, and enters the second-stage impeller. This process is repeated until the fully compressed air leaves through the final pinion-impeller, or stage.

Most bullgear compressors are designed to operate with a gear speed of 3,600 rpm. In a typical four-stage compressor, the pinions operate at progressively higher speeds. A typical range is between 12,000 rpm (first stage) and 70,000 rpm (fourth stage).

Due to their cantilever design and pinion rotating speeds, bullgear compressors are extremely sensitive to variations in demand or downstream pressure changes. Because of this sensitivity, their use should be limited to baseload applications.

Bullgear compressors are not designed for, nor will they tolerate, load-following applications. They should not be installed in the same discharge manifold with posi-

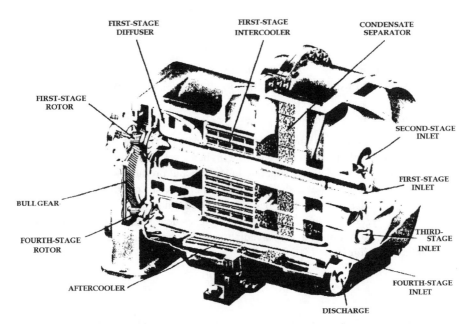

Figure 10–4 Bullgear centrifugal compressor (Gibbs 1971).

tive-displacement compressors, especially reciprocating compressors. The standing-wave pulses created by many positive-displacement compressors create enough variation in the discharge manifold to cause potentially serious instability.

In addition, the large helical gear used for the bullgear creates an axial oscillation or thrusting that contributes to instability within the compressor. This axial movement is transmitted throughout the machine train.

Performance

Compressed-air systems and compressors are governed by the physical laws of thermodynamics, which define their efficiency and system dynamics. This section discusses the first and second laws of thermodynamics, which apply to all compressors and compressed-air systems. Also applying to these systems are the ideal gas law and the concepts of pressure and compression.

First Law of Thermodynamics

This law states that energy cannot be created or destroyed during a process, such as compression and delivery of air or gas, although it may change from one form of energy to another. In other words, whenever a quantity of one kind of energy disappears, an exactly equivalent total of other kinds of energy must be produced. This is expressed for a steady-flow open system such as a compressor by the following relationship:

| Net energy added to system as heat and work | + | Stored energy of mass entering system | − | Stored energy of mass leaving system | = 0 |

Second Law of Thermodynamics

The second law of thermodynamics states that energy exists at various levels and is available for use only if it can move from a higher level to a lower one. For example, it is impossible for any device to operate in a cycle and produce work while exchanging heat only with bodies at a single, fixed temperature. In thermodynamics, a measure of the unavailability of energy has been devised, known as *entropy*. As a measure of unavailability, entropy increases as a system loses heat, but remains constant when there is no gain or loss of heat as in an adiabatic process. It is defined by the following differential equation:

$$dS = \frac{dQ}{T}$$

where T is the temperature (Fahrenheit) and Q is the heat added (BTU).

Pressure/Volume/Temperature Relationship

Pressure (P), temperature (T), and volume (V) are properties of gases that are completely interrelated. Boyle's law and Charles' law may be combined into one equation that is referred to as the *ideal gas law*. This equation is true always for ideal gases and for real gases under certain conditions:

$$\frac{P_1 V_1}{T_1} = \frac{P_2 V_2}{T_2}$$

For air at room temperature, the error in this equation is less than 1 percent for pressures as high as 400 psia (absolute psi). For air at one atmosphere of pressure, the error is less than 1 percent for temperatures as low as −200° Fahrenheit. These error factors will vary for different gases.

Pressure/Compression

In a compressor, pressure is generated by pumping quantities of gas into a tank or other pressure vessel. The pressure is increased by progressively increasing the amount of gas in the confined or fixed-volume space. The effects of pressure exerted by a confined gas result from the force acting on the container walls. This force is caused by the rapid and repeated bombardment from the enormous number of molecules present in a given quantity of gas.

Compression occurs when the space between the molecules is decreased. Less volume means that each particle has a shorter distance to travel, thus proportionately more collisions occur in a given span of time, resulting in a higher pressure. Air compressors are designed to generate particular pressures to meet specific application requirements.

Other Performance Indicators

Centrifugal compressors are governed by the same performance indicators as centrifugal pumps or fans.

Installation

Dynamic compressors seldom pose serious foundation problems. Since moments and shaking forces are not generated during compressor operation, there are no variable loads to be supported by the foundation. A foundation or mounting of sufficient area and mass to maintain the compressor level and alignment and to assure safe soil loading is all that is required. The units may be supported on structural steel if necessary. The principles defined in the section on Operating Dynamics in Chapter 8 for centrifugal pumps also apply to centrifugal compressors.

It is necessary to install pressure-relief valves on most dynamic compressors to protect them, due to restrictions placed on casing pressure, power input, and to keep out of its surge range. Always install a valve capable of bypassing the full-load capacity of the compressor between its discharge port and the first isolation valve.

Operating Methods

The acceptable operating envelope for centrifugal compressors is very limited. Therefore, care should be taken to minimize any variation in suction supply, back-pressure caused by changes in demand, and frequency of unloading. The operating guidelines provided in the compressor vendor's O&M manual should be followed to prevent abnormal operating behavior or premature wear or failure of the system.

Centrifugal compressors are designed to be baseloaded and may exhibit abnormal behavior or chronic reliability problems when used in a load-following mode of operation. This is especially true of bullgear and cantilever compressors. For example, a 1-psig change in discharge pressure may be enough to cause catastrophic failure of a bullgear compressor. Variations in demand or back pressure on a cantilever design can cause the entire rotating element and its shaft to flex. This not only affects the compressor's efficiency but also accelerates wear and may lead to premature shaft or rotor failure.

All compressor types have moving parts, high noise levels, high pressures, and high-temperature cylinder and discharge-piping surfaces.

POSITIVE DISPLACEMENT

Positive-displacement compressors can be divided into two major classifications: rotary and reciprocating.

Rotary

The rotary compressor is adaptable to direct drive by the use of induction motors or multicylinder gasoline or diesel engines. These compressors are compact, relatively inexpensive, and require a minimum of operating attention and maintenance. They occupy a fraction of the space and weight of a reciprocating machine having equivalent capacity.

Rotary compressors are classified into three general groups: sliding vane, helical lobe, and liquid-seal ring.

Sliding Vane

The basic element of the sliding-vane compressor is the cylindrical housing and the rotor assembly. This compressor, illustrated in Figure 10–5, has longitudinal vanes that slide radially in a slotted rotor mounted eccentrically in a cylinder. The centrifugal force carries the sliding vanes against the cylindrical case with the vanes forming a number of individual longitudinal cells in the eccentric annulus between the case and rotor. The suction port is located where the longitudinal cells are largest. The size of each cell is reduced by the eccentricity of the rotor as the vanes approach the discharge port, thus compressing the gas.

Cyclical opening and closing of the inlet and discharge ports occurs by the rotor's vanes passing over them. The inlet port normally is a wide opening designed to admit gas in the pocket between two vanes. The port closes momentarily when the second vane of each air-containing pocket passes over the inlet port.

When running at design pressure, the theoretical operation curves (see Figure 10–6) are identical to a reciprocating compressor. However, there is one major difference between a sliding-vane and a reciprocating compressor. The reciprocating unit has spring-loaded valves that open automatically with small pressure

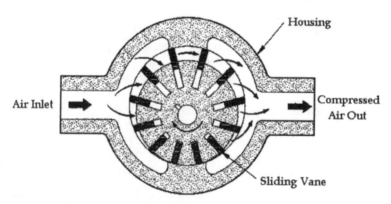

Figure 10–5 Rotary sliding-vane compressor (Gibbs 1971).

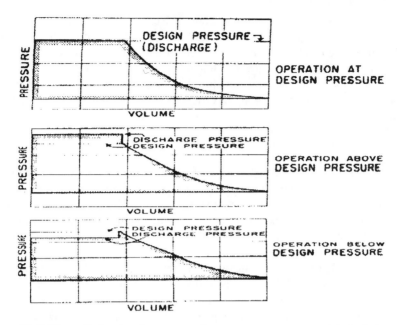

Figure 10–6 Theoretical operation curves for rotary compressors with built-in porting (Gibbs 1971).

differentials between the outside and inside cylinder. The sliding-vane compressor has no valves.

The fundamental design considerations of a sliding-vane compressor are the rotor assembly, the cylinder housing, and the lubrication system.

Housing and Rotor Assembly   Cast iron is the standard material used to construct the cylindrical housing, but other materials may be used if corrosive conditions exist. The rotor usually is a continuous piece of steel that includes the shaft and is made from bar stock. Special materials can be selected for corrosive applications. Occasionally, the rotor may be a separate iron casting keyed to a shaft. On most standard air compressors, the rotor-shaft seals are semi-metallic packing in a stuffing box. Commercial mechanical rotary seals can be supplied when needed. Cylindrical roller bearings generally are used in these assemblies.

Vanes usually are asbestos or cotton cloth impregnated with a phenolic resin. Bronze or aluminum also may be used for vane construction. Each vane fits into a milled slot extending the full length of the rotor and slides radially in and out of this slot once per revolution. Vanes are the part in the compressor most in need of maintenance. Each rotor has from 8 to 20 vanes, depending on its diameter. A greater number of vanes increases compartmentalization, which reduces the pressure differential across each vane.

Lubrication System A V-belt-driven, force-fed oil lubrication system is used on water-cooled compressors. Oil goes to both bearings and several points in the cylinder. Ten times as much oil is recommended to lubricate the rotary cylinder as is required for the cylinder of a corresponding reciprocating compressor. The oil carried over with the gas to the line may be reduced 50 percent with an oil separator on the discharge. Use of an aftercooler ahead of the separator permits removal of 85 to 90 percent of the entrained oil.

Helical Lobe or Screw

The helical lobe, or screw, compressor is shown in Figure 10–7. It has two or more mating sets of lobe-type rotors mounted in a common housing. The male lobe, or rotor, usually is driven directly by an electric motor. The female lobe, or mating rotor, is driven by a helical gear set mounted on the outboard end of the rotor shafts. The gears provide both motive power for the female rotor and absolute timing between the rotors.

The rotor set has extremely close mating clearance (i.e., about 0.5 mils) but no metal-to-metal contact. Most of these compressors are designed for oil-free operation. In other words, no oil is used to lubricate or seal the rotors. Instead, oil lubrication is limited to the timing gears and bearings outside the air chamber. Because of this, maintaining proper clearance between the two rotors is critical.

This type of compressor is classified as a constant volume, variable-pressure machine that is quite similar to the vane-type rotary in general characteristics. Both have a built-in compression ratio.

Helical-lobe compressors are best suited for baseload applications, where they can provide a constant volume and pressure of discharge gas. The only recommended method of volume control is the use of variable-speed motors. With variable-speed

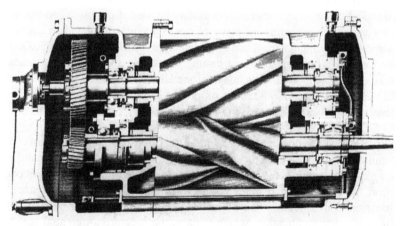

Figure 10–7 Helical lobe, or screw, rotary air compressor (Gibbs 1971).

drives, capacity variations can be obtained with a proportionate reduction in speed. A 50 percent speed reduction is the maximum permissible control range.

Helical-lobe compressors are not designed for frequent or constant cycles between load and no-load operation. Each time the compressor unloads, the rotors tend to thrust axially. Even though the rotors have a substantial thrust bearing and, in some cases, a balancing piston to counteract axial thrust, the axial clearance increases each time the compressor unloads. Over time, this clearance will increase enough to permit a dramatic rise in the impact energy created by axial thrust during the transition from loaded to unloaded conditions. In extreme cases, the energy can be enough to physically push the rotor assembly through the compressor housing.

The compression ratio and maximum inlet temperature determine the maximum discharge temperature of these compressors. Discharge temperatures must be limited to prevent excessive distortion between the inlet and discharge ends of the casing and rotor expansion. High-pressure units are water-jacketed to obtain uniform casing temperature. Rotors also may be cooled to permit a higher operating temperature.

Either casing distortion or rotor expansion can cause the clearance between the rotating parts to decrease and allow metal-to-metal contact. Since the rotors typically rotate at speeds between 3,600 and 10,000 rpm, metal-to-metal contact normally results in instantaneous, catastrophic compressor failure.

Changes in differential pressures can be caused by variations in either inlet or discharge conditions (i.e., temperature, volume, or pressure). Such changes can cause the rotors to become unstable and change the load zones in the shaft-support bearings. The result is premature wear or failure of the bearings.

Always install a relief valve capable of bypassing the full-load capacity of the compressor between its discharge port and the first isolation valve. Since helical-lobe compressors are less tolerant to overpressure operation, safety valves usually are set within 10 percent of absolute discharge pressure, or 5 psi, whichever is lower.

Liquid-Seal Ring

The liquid-ring, or liquid-piston, compressor is shown in Figure 10–8. It has a rotor with multiple forward-turned blades rotating about a central cone that contains inlet and discharge ports. Liquid is trapped between adjacent blades, which drive the liquid around the inside of an elliptical casing. As the rotor turns, the liquid face moves in and out of this space due to the casing shape, creating a liquid piston. Porting in the central cone is built in and fixed, and there are no valves.

Compression occurs within the pockets or chambers between the blades before the discharge port is uncovered. Since the port location must be designed and built for a specific compression ratio, it tends to operate above or below the design pressure (refer back to Figure 10–6).

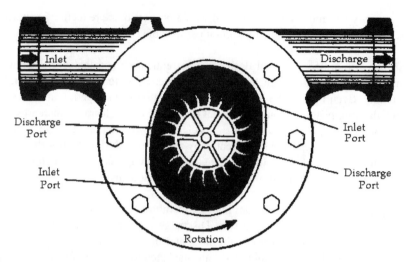

Figure 10–8 Liquid-seal ring rotary air compressor (Gibbs 1971).

Liquid-ring compressors are cooled directly rather than by jacketed casing walls. The cooling liquid is fed into the casing where it comes into direct contact with the gas being compressed. The excess liquid is discharged with the gas. The discharged mixture is passed through a conventional baffle or centrifugal-type separator to remove the free liquid. Because of the intimate contact of gas and liquid, the final discharge temperature can be held close to the inlet cooling water temperature. However, the discharge gas is saturated with liquid at the discharge temperature of the liquid.

The amount of liquid passed through the compressor is not critical and can be varied to obtain the desired results. The unit will not be damaged if a large quantity of liquid inadvertently enters its suction port.

Lubrication is required only in the bearings, which generally are located external to the casing. The liquid itself acts as a lubricant, sealing medium, and coolant for the stuffing boxes.

Performance

Performance of a rotary positive-displacement compressor can be evaluated using the same criteria as a positive-displacement pump. Also refer to the previous discussion of the laws of thermodynamics that apply to all compressors. As constant-volume machines, performance is determined by rotation speed, internal slip, and total back pressure on the compressor.

The volumetric output of rotary positive-displacement compressors can be controlled by changing the speed. The slower the compressor turns, the lower is its output volume. This feature permits the use of these compressors in load-following applications. However, care must be taken to prevent sudden, radical changes in speed.

Internal slip is simply the amount of gas that can flow through internal clearances from the discharge back to the inlet. Obviously, internal wear will increase internal slip.

Discharge pressure is relatively constant regardless of operating speed. With the exception of slight pressure variations caused by atmospheric changes and back pressure, a rotary positive-displacement compressor will provide a fixed discharge pressure. Back pressure, which is caused by restrictions in the discharge piping or demand from users of the compressed air or gas, can have a serious impact on compressor performance.

If back pressure is too low or demand too high, the compressor will be unable to provide sufficient volume or pressure to the downstream systems. In this instance, the discharge pressure will be noticeably lower than designed.

If back pressure is too high or demand too low, the compressor will generate a discharge pressure higher than designed. It will continue to compress the air or gas until it reaches the unload setting on the system's relief valve or until the brake horsepower required exceeds the maximum horsepower rating of the driver.

Installation

Installation requirements for rotary positive-displacement compressors are similar to any rotating machine. As with centrifugal compressors, rotary positive-displacement compressors must be fitted with pressure-relief devices to limit the discharge or interstage pressures to a safe maximum for the equipment served.

In applications where demand varies, rotary positive-displacement compressors require a downstream receiver tank or reservoir that minimizes the load-unload cycling frequency of the compressor. The receiver tank should have sufficient volume to permit acceptable unload frequencies for the compressor. Refer to the vendor's O&M manual for specific receiver-tank recommendations.

Operating Methods

All compressor types have moving parts, high noise levels, high pressures, and high-temperature cylinder and discharge-piping surfaces.

Rotary positive-displacement compressors should be operated as baseloaded units. They are especially sensitive to the repeated start-stop operation required by load-following applications. Generally, rotary positive-displacement compressors are designed to unload about every six to eight hours. This unload cycle is needed to dissipate the heat generated by the compression process. If the unload frequency is too great, these compressors have a high probability of failure.

The primary operating control inputs for rotary positive-displacement compressors are discharge pressure, pressure fluctuation, and unloading frequency.

Discharge Pressure This type of compressor will continue to compress the air volume in the downstream system until (1) some component in the system fails, (2) the brake horsepower exceeds the driver's capacity, or (3) a safety valve opens. Therefore, the operator's primary control input should be the compressor's discharge pressure. A discharge pressure below the design point is a clear indicator that the total downstream demand is greater than the unit's capacity. If the discharge pressure is too high, the demand is too low and excessive unloading will be required to prevent failure.

Pressure Fluctuation Fluctuations in the inlet and discharge pressures indicate potential system problems that may adversely affect performance and reliability. Pressure fluctuation generally is caused by changes in the ambient environment, turbulent flow, or restrictions due to partially blocked inlet filters. Any of these problems will result in performance and reliability problems if not corrected.

Unloading Frequency The unloading function in rotary positive-displacement compressors is automatic and not under operator control. Generally, a set of limit switches, one monitoring internal temperature and one monitoring discharge pressure, is used to trigger the unloading process. By design, the limit switch that monitors the compressor's internal temperature is the primary control. The secondary control, or discharge-pressure switch, is a fail-safe design to prevent overloading the compressor.

Depending on design, rotary positive-displacement compressors have an internal mechanism designed to minimize the axial thrust caused by the instantaneous change from fully loaded to unloaded operating conditions. In some designs, a balancing piston is used to absorb the rotor's thrust during this transition. In others, oversized thrust bearings are used.

Regardless of the mechanism used, none provides complete protection from the damage imparted by the transition from load to no-load conditions. However, as long as the unloading frequency is within design limits, this damage will not adversely affect the compressor's useful operating life or reliability. However, an unloading frequency greater than that accommodated in the design will reduce the useful life of the compressor and may lead to premature, catastrophic failure.

Operating practices should minimize, as much as possible, the unloading frequency of these compressors. Installation of a receiver tank and modification of user-demand practices are the most effective solutions to this type of problem.

Reciprocating

Reciprocating compressors are widely used by industry and are offered in a wide range of sizes and types. They vary from units requiring less than 1 hp to more than

12,000 hp. Pressure capabilities range from low vacuums at intake to special compressors capable of 60,000 psig or higher.

Reciprocating compressors are classified as constant-volume, variable-pressure machines. They are the most efficient type of compressor and can be used for partial-load, or reduced-capacity, applications.

Because of the reciprocating pistons and unbalanced rotating parts, the unit tends to shake. Therefore, it is necessary to provide a mounting that stabilizes the installation. The extent of this requirement depends on the type and size of the compressor.

Reciprocating compressors should be supplied with clean gas, so inlet filters are recommended in all applications. The filters cannot satisfactorily handle liquids entrained in the gas, although vapors are no problem if condensation within the cylinders does not take place. Liquids will destroy the lubrication and cause excessive wear.

Reciprocating compressors deliver a pulsating flow of gas that can damage downstream equipment or machinery. This sometimes is a disadvantage, but pulsation dampers can be used to alleviate the problem.

Configuration

Certain design fundamentals should be clearly understood before analyzing the operating condition of reciprocating compressors. These fundamentals include frame and running gear, inlet and discharge valves, cylinder cooling, and cylinder orientation.

Frame and Running Gear Two basic factors guide frame and running gear design. The first factor is the maximum horsepower to be transmitted through the shaft and running gear to the cylinder pistons. The second factor is the load imposed on the frame parts by the pressure differential between the two sides of each piston. This often is called *pin load* because the full force is exerted directly on the crosshead and crankpin. These two factors determine the size of bearings, connecting rods, frame, and bolts that must be used throughout the compressor and its support structure.

Cylinder Design Compression efficiency depends entirely on the design of the cylinder and its valves. Unless the valve area is sufficient to allow gas to enter and leave the cylinder without undue restriction, efficiency cannot be high. Valve placement for free flow of the gas in and out of the cylinder also is important.

Both efficiency and maintenance are influenced by the degree of cooling during compression. The method of cylinder cooling must be consistent with the service intended.

The cylinders and all the parts must be designed to withstand the maximum application pressure. The most economical materials that will give the proper strength and the longest service under the design conditions generally are used.

Inlet and Discharge Valves   Compressor valves are placed in each cylinder to per-mit one-way flow of gas, either into or out of the cylinder. One or more valve(s) is needed for inlet and discharge in each compression chamber.

Each valve opens and closes once for each revolution of the crankshaft. The valves in a compressor operating at 700 rpm for 8 hours per day and 250 days per year will have cycled (i.e., opened and closed) 42,000 times per hour, 336,000 times per day, or 84 million times in a year. The valves have less than $\frac{1}{10}$ of a second to open, let the gas pass through, and close.

They must cycle with a minimum of resistance for minimum power consumption. However, the valves must have minimal clearance to prevent excessive expansion and reduced volumetric efficiency. They must be tight under extreme pressure and temper-ature conditions. Finally, the valves must be durable under many kinds of abuse.

Four basic valve designs are used in these compressors: finger, channel, leaf, and annular ring. Each class may contain variations in design, depending on operating speed and size of valve required.

- *Finger.* Figure 10–9 is an exploded view of a typical finger valve. These valves are used for smaller, air-cooled compressors. One end of the finger is fixed, and the opposite end lifts when the valve opens.
- *Channel.* The channel valve shown in Figure 10–10 is widely used in mid-to large-sized compressors. This valve uses a series of separate stainless

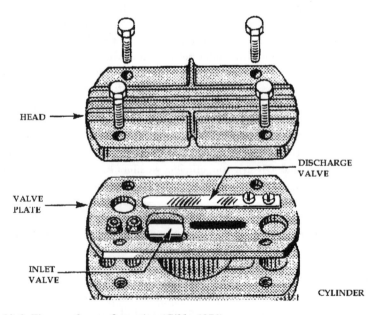

Figure 10–9 Finger valve configuration (Gibbs 1971).

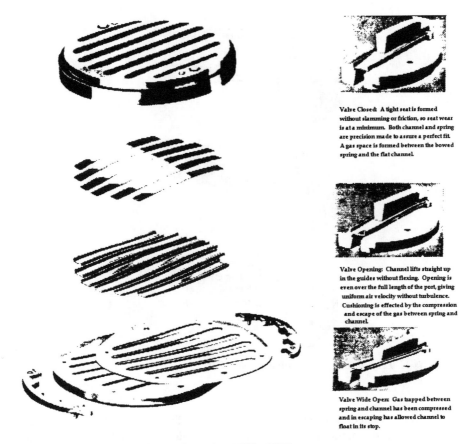

Valve Closed: A tight seat is formed without slamming or friction, so seat wear is at a minimum. Both channel and spring are precision made to assure a perfect fit. A gas space is formed between the bowed spring and the flat channel.

Valve Opening: Channel lifts straight up in the guides without flexing. Opening is even over the full length of the port, giving uniform air velocity without turbulence. Cushioning is effected by the compression and escape of the gas between spring and channel.

Valve Wide Open: Gas trapped between spring and channel has been compressed and in escaping has allowed channel to float in its stop.

Figure 10–10 Channel valve configuration (Gibbs 1971).

steel channels. As explained in the figure, this is a cushioned valve, which adds greatly to its life.

- *Leaf.* The leaf valve (see Figure 10–11) has a configuration somewhat like the channel valve. It is made of flat-strip steel that opens against an arched stop plate. This results in valve flexing only at its center with maximum lift. The valve operates as its own spring.
- *Annular Ring.* Figure 10–12 shows exploded views of typical inlet and discharge annular-ring valves. The valves shown have a single ring, but larger sizes may have two or three rings. In some designs, the concentric rings are tied into a single piece by bridges. The springs and valve move into a recess in the stop plate as the valve opens. Gas trapped in the recess acts as a cushion and prevents slamming. This eliminates a major source of valve and spring breakage. The valve shown was the first cushioned valve built.

Cylinder Cooling Cylinder heat is produced by the work of compression plus friction caused by the action of the piston and piston rings on the cylinder wall and packing on

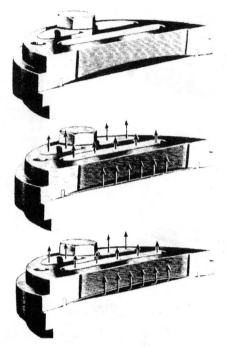

Figure 10–11 Leaf spring configuration (Gibbs 1971).

the rod. The amount of heat generated can be considerable, particularly when moderate to high compression ratios are involved. This can result in undesirably high operating temperatures.

Most compressors use some method to dissipate a portion of this heat to reduce the cylinder wall and discharge gas temperatures. The following are advantages of cylinder cooling:

- Lowering cylinder wall and cylinder head temperatures reduces loss of capacity and horsepower per unit volume due to suction gas preheating during inlet stroke. This results in more gas in the cylinder for compression.
- Reducing cylinder wall and cylinder head temperatures removes more heat from the gas during compression, lowering its final temperature and reducing the power required.
- Reducing the gas temperature and that of the metal surrounding the valves results in a longer valve service life and reduces the possibility of deposit formation.
- Reduced cylinder wall temperature promotes better lubrication, resulting in longer life and reduced maintenance.
- Cooling, particularly water cooling, maintains a more even temperature around the cylinder bore and reduces warpage.

Figure 10–12 Annular-ring valves (Gibbs 1971).

Cylinder Orientation   Orientation of the cylinders in a multistage or multicylinder compressor directly affects the operating dynamics and vibration level. Figure 10–13 illustrates a typical three-piston, air-cooled compressor. Since three pistons are oriented within a 120° arc, this type of compressor generates higher vibration levels than the opposed piston compressor illustrated in Figure 10–14.

Performance

Refer back to the previous discussion of the laws of thermodynamics governing all compressors. Reciprocating-compressor performance is governed almost exclusively by operating speed. Each cylinder of the compressor will discharge the same volume, excluding slight variations caused by atmospheric changes, at the same discharge pressure each time it completes the discharge stroke. As the rotation speed of the compressor changes, so does the discharge volume.

The only other variables that affect performance are the inlet-discharge valves, which control flow into and out of each cylinder. Although reciprocating compressors can use a variety of valve designs, it is crucial that the valves perform reliably. If they are damaged and fail to operate at the proper time or do not seal properly, overall compressor performance will be reduced substantially.

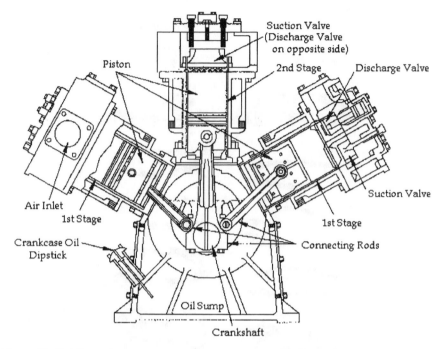

Figure 10–13 Three-piston compressor generates higher vibration levels (Gibbs 1971).

Installation

A carefully planned and executed installation is extremely important and makes compressor operation and maintenance easier and safer. Key components of a compressor installation are location, foundation, and piping.

Location   The preferred location for any compressor is near the center of its load. However, the choice often is influenced by the cost of supervision, which can vary by location. The ongoing cost of supervision may be less expensive at a less-optimum location, which can offset the cost of longer piping.

A compressor always will give better, more reliable service when enclosed in a building that protects it from cold, dusty, damp, and corrosive conditions. In certain locations, it may be economical to use a roof only, but this is not recommended unless the weather is extremely mild. Even then, it is crucial to prevent rain and wind-blown debris from entering the moving parts. Subjecting a compressor to adverse inlet conditions will dramatically reduce its reliability and significantly increase its maintenance requirements.

Ventilation around a compressor is vital. On a motor-driven, air-cooled unit, the heat radiated to the surrounding air is at least 65 percent of the power input. On a water-

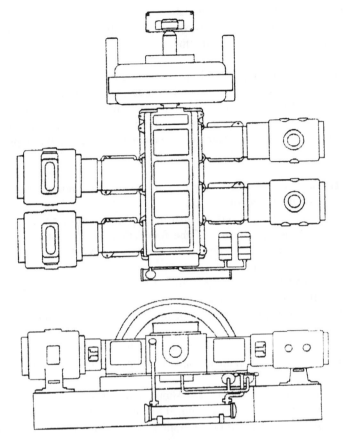

Figure 10–14 Opposed-piston compressor balances piston forces.

jacketed unit with an aftercooler and outside receiver, the heat radiated to the surrounding air may be 15 to 25 percent of the total energy input, which still is a substantial amount of heat. Positive outside ventilation is recommended for any compressor room where the ambient temperature may exceed 104°F.

Foundation Because of the alternating movement of pistons and other components, reciprocating compressors often develop a shaking that alternates in direction. This force must be damped and contained by the mounting. The foundation also must support the weight load of the compressor and its driver.

There are many compressor arrangements and the net magnitude of the moments and forces developed can vary a great deal among them. In some cases, they are partially or completely balanced within the compressors themselves. In others, the foundation must handle much of the force. When complete balance is possible, reciprocating compressors can be mounted on a foundation just large and rigid enough to carry the weight and maintain alignment. However, most reciprocating compressors require larger, more massive foundations than other machinery.

Depending on the size and type of unit, the mounting may vary from simply bolting it to the floor to attaching to a massive foundation designed specifically for the application. A proper foundation must (1) maintain the alignment and level of the compressor and its driver at the proper elevation and (2) minimize vibration and prevent its transmission to adjacent building structures and machinery. There are five steps to accomplish the first objective:

1. The safe weight-bearing capacity of the soil must not be exceeded at any point on the foundation base.
2. The load to the soil must be distributed over the entire area.
3. The size and proportion of the foundation block must be such that the resultant vertical load due to the compressor, block, and any unbalanced force falls within the base area.
4. The foundation must have sufficient mass and weight-bearing area to prevent its sliding on the soil due to unbalanced forces.
5. Foundation temperature must be uniform to prevent warping.

Bulk is not usually the complete solution to foundation problems. A certain weight sometimes is necessary, but soil area usually is of more value than foundation mass.

Determining if two or more compressors should have separate or single foundations depends on the compressor type. A combined foundation is recommended for reciprocating units, since the forces from one unit usually will partially balance out the forces from the others. In addition, the greater mass and surface area in contact with the ground damps foundation movement and provides greater stability.

Soil quality may vary seasonally, and such conditions must be carefully considered in the foundation design. No foundation should rest partially on bedrock and partially on soil; it should rest entirely on one or the other. If placed on the ground, make sure that part of the foundation does not rest on soil that has been disturbed. In addition, pilings may be necessary to ensure stability.

Piping   Piping should easily fit the compressor connections without needing to spring or twist it to fit. It must be supported independently of the compressor and anchored, as necessary, to limit vibration and to prevent expansion strains. Improperly installed piping may distort or pull the compressor's cylinders or casing out of alignment.

Air Inlet   The intake pipe on an air compressor should be as short and direct as possible. If the total run of the inlet piping is unavoidably long, the diameter should be increased. The pipe size should be greater than the compressor's air-inlet connection.

Cool inlet air is desirable. For every 5°F of ambient air temperature reduction, the volume of compressed air generated increases by 1 percent with the same power consumption. This increase in performance is due to the greater density of the intake air.

It is preferable for the intake air to be taken from outdoors. This reduces heating and air conditioning costs and, if properly designed, has fewer contaminants. However, the intake piping should be a minimum of 6 ft above the ground and screened or, preferably, filtered. An air inlet must be free of steam and engine exhaust. The inlet should be hooded or turned down to prevent the entry of rain or snow. It should be above the building eaves and several feet from the building.

Discharge   Discharge piping should be the full size of the compressor's discharge connection. The pipe size should not be reduced until the point along the pipeline is reached where the flow has become steady and nonpulsating. With a reciprocating compressor, this generally is beyond the aftercooler or the receiver. Pipes to handle nonpulsating flow are sized by normal methods, and long-radius bends are recommended. All discharge piping must be designed to allow adequate expansion loops or bends to prevent undue stress at the compressor.

Drainage   Before piping is installed, the layout should be analyzed to eliminate low points where liquid could collect and to provide drains where low points cannot be eliminated. A regular part of the operating procedure must be the periodic drainage of low points in the piping and separators, as well as inspection of automatic drain traps.

Pressure-Relief Valves   All reciprocating compressors must be fitted with pressure-relief devices to limit the discharge or interstage pressures to a safe maximum for the equipment served. Always install a relief valve capable of bypassing the full-load capacity of the compressor between its discharge port and the first isolation valve. The safety valves should be set to open at a pressure slightly higher than the normal discharge-pressure rating of the compressor. For standard 100 to 115 psig two-stage air compressors, safety valves normally are set at 125 psig.

The pressure-relief safety valve normally is situated on top of the air reservoir, and there must be no restriction on its operation. The valve usually is of the "huddling chamber" design, in which the static pressure acting on its disk area causes it to open. Figure 10–15 illustrates how such a valve functions. As the valve pops, the air space within the huddling chamber between the seat and blowdown ring fills with pressurized air and builds up more pressure on the roof of the disk holder. This temporary pressure increases the upward thrust against the spring, causing the disk and its holder to fully pop open.

Once a predetermined pressure drop (i.e., blowdown) occurs, the valve closes with a positive action by trapping pressurized air on top of the disk holder. The pressure-drop setpoint is adjusted by raising or lowering the blowdown ring. Raising the ring increases the pressure-drop setting, while lowering it decreases the setting.

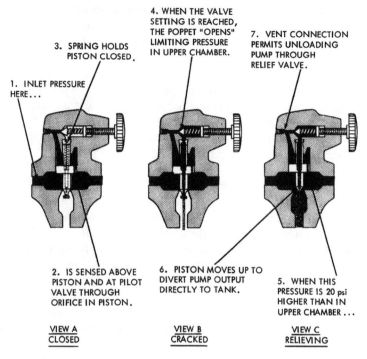

Figure 10–15 Illustrates how a safety valve functions.

Operating Methods

Compressors can be hazardous to work around because they have moving parts. Ensure that clothing is kept away from belt drives, couplings, and exposed shafts. In addition, high-temperature surfaces around cylinders and discharge piping are exposed. Compressors are notoriously noisy, so ear protection should be worn. These machines are used to generate high-pressure gas so, when working around them, it is important to wear safety glasses and to avoid searching for leaks with bare hands. High-pressure leaks can cause severe friction burns.

11

MIXERS AND AGITATORS

Mixers are devices that blend combinations of liquids and solids into a homogenous product. They come in a variety of sizes and configurations designed for specific applications. Agitators provide the mechanical action to keep dissolved or suspended solids in solution.

Both operate on basically the same principles, but variations in design, operating speed, and applications divide the actual function of these devices. Agitators generally work just as hard as mixers, and the terms often are used interchangeably.

CONFIGURATION

There are two primary types of mixers: propeller/paddle and screw. Screw mixers can be further divided into batch and mixer-extruder types.

Propeller/Paddle

Propeller/paddle mixers are used to blend or agitate liquid mixtures in tanks, pipelines, or vessels. Figure 11–1 illustrates a typical top-entering propeller/paddle mixer. This unit consists of an electric motor, a mounting bracket, an extended shaft, and one or more impeller(s) or propeller(s). Materials of construction range from bronze to stainless steel, which are selected based on the particular requirements of the application.

The propeller/paddle mixer also is available in a side-entering configuration, which is shown in Figure 11–2. This configuration typically is used to agitate liquids in large vessels or pipelines. The side-entering mixer is essentially the same as the top-entering version except for the mounting configuration.

Both the top-entering and side-entering mixers may use either propellers, as shown in the preceding figures, or paddles, as illustrated by Part b of Figure 11–3. Generally,

Figure 11–1 Top-entering propeller-type mixer (Thomas Register 1995).

Figure 11–2 Side-entering propeller-type mixer (Thomas Register 1995).

Figure 11–3 Mixer can use either propellers or paddles to provide agitation (Thomas Register 1995).

propellers are used for medium- to high-speed applications where the viscosity is relatively low. Paddles are used in low-speed, high-viscosity applications.

Screw

The screw mixer uses a single- or dual-screw arrangement to mix liquids, solids, or a combination of both. It comes in two basic configurations: batch and combination mixer-extruder.

Batch

Figure 11–4 illustrates a typical batch-type screw mixer. This unit consists of a mixing drum or cylinder, a single- or dual-screw mixer, and a power supply.

The screw configuration normally is either a ribbon-type helical screw or a series of paddles mounted on a common shaft. Materials of construction are selected based on the specific application and materials to be mixed. Typically, the screws are either steel or stainless steel, but other materials are available.

Combination Mixer-Extruder

The mixer-extruder combination unit shown in Figure 11–5 combines the functions of a mixer and a screw conveyor. This type of mixer is used for mixing viscous products.

Figure 11–4 Batch-type mixer uses single or dual screws to mix product (Thomas Register 1995).

PERFORMANCE

Unlike centrifugal pumps and compressors, few criteria can be used directly to determine mixer effectiveness and efficiency. However, the product quality and brake horsepower are indices that can be used to indirectly gauge performance.

Product Quality

The primary indicator of acceptable performance is the quality of the product delivered by the mixer. Although there is no direct way to measure this indicator, feedback from the quality assurance group should be used to verify that acceptable performance levels are attained.

Brake Horsepower

Variation in the actual brake horsepower required to operate a mixer is the primary indicator of its performance envelope. Mixer design, whether propeller or screw type, is based on the viscosity of both the incoming and finished product. These variables determine the brake horsepower required to drive the mixer, which will follow varia-

Figure 11–5 Combination mixer-extruder (Thomas Register 1995).

tions in the viscosity of the products being mixed. As the viscosity increases so will the brake horsepower demand. Conversely, as the viscosity decreases, so will the horsepower required to drive the mixer.

INSTALLATION

Installation of propeller-type mixers varies greatly, depending on the specific application. Top-entering mixers utilize either a clamp- or flange-type mounting. It is important that the mixer be installed so the propeller or paddle is at a point within the tank, vessel, or piping that assures proper mixing. Vendor recommendations found in O&M manuals should be followed to ensure proper operation of the mixer.

Mixers should be mounted on a rigid base that assures level alignment and prevents lateral movement of the mixer and its drivetrain. While most mixers can be bolted directly to a base, care must be taken to ensure that the base is rigid and has the structural capacity to stabilize the mixer.

OPERATING METHODS

There are only three major operating concerns for mixers: setup, incoming-feed rate, and product viscosity.

Mixer Setup

Both propeller and screw mixers have specific setup requirements. In the case of propeller/paddle-type mixers, the primary factor is the position of the propellers or paddles within the tank or vessel. Vendor recommendations should be followed to assure proper operation of the mixer.

If the propellers or paddles are too close to the liquid level, the mixer will create a vortex that will entrain air and prevent adequate blending or mixing. If the propellers are set too low, compress vortexing may occur. When this happens, the mixer will create a stagnant zone in the area under the rotating assembly. As a result, some of the product will settle in this zone and proper mixing cannot occur. Setting the mixer too close to a corner or the side of the mixing vessel also can create a stagnant zone that will prevent proper blending or mixing of the product.

For screw-type mixers, proper clearance between the rotating element and the mixer housing must be maintained to vendor specifications. If the clearance is improperly set, the mixer will bind (i.e., not enough clearance) or fail to blend properly.

Feed Rate

Mixers are designed to handle a relatively narrow band of incoming product flow rate. Therefore, care must be exercised to ensure that the actual feed rate is maintained within acceptable limits. The O&M manuals provided by the vendor will provide the feed-rate limitations for various products. Normally, these rates must be adjusted for viscosity and temperature variation.

Viscosity

Variations in viscosity of both the incoming and finished products have a dramatic effect on mixer performance. Standard operating procedures should include specific operating guidelines for the range of variation acceptable for each application. The recommended range should include adjustments for temperature, flow rate, mixing speed, and other factors that directly or indirectly affect viscosity.

12

DUST COLLECTORS

The basic operations performed by dust-collection devices are (1) separating particles from the gas stream by deposition on a collection surface, (2) retaining the deposited particles on the surface until removal, and (3) removing the deposit from the surface for recovery or disposal.

The separation step requires (1) application of a force that produces a differential motion of the particles relative to the gas and (2) sufficient gas-retention time for the particles to migrate to the collecting surface. Most dust-collections systems are constituted of a pneumatic-conveying system and some device that separates suspended particulate matter from the conveyed airstream. The more common systems use either filter media (e.g., fabric bags) or cyclonic separators to separate the particulate matter from air.

BAGHOUSES

Fabric-filter systems, commonly called *bag-filter* or *baghouse systems*, are dust-collection systems in which dust-laden air is passed through a bag-type filter. The bag collects the dust in layers on its surface and the dust layer itself effectively becomes the filter medium. Because the bag's pores usually are much larger than those of the dust-particle layer that forms, the initial efficiency is very low. However, efficiency improves once an adequate dust layer forms. Therefore, the potential for dust penetration of the filter media is extremely low except during the initial period after startup, bag change, or during the fabric-cleaning, or blow-down, cycle.

The principal mechanisms of disposition in dust collectors are (1) gravitational deposition, (2) flow-line interception, (3) inertial deposition, (4) diffusional deposition, and (5) electrostatic deposition. During the initial operating period, particle deposition takes place mainly by inertial and flow-line interception, diffusion, and gravity.

Once the dust layer has been fully established, sieving probably is the dominant deposition mechanism.

Configuration

A baghouse system consists of the following: a pneumatic-conveyor system, filter media, a back-flush cleaning system, and a fan or blower to provide airflow.

Pneumatic Conveyor

The primary mechanism for conveying dust-laden air to a central collection point is a system of pipes or ductwork that functions as a pneumatic conveyor. This system gathers dust-laden air from various sources within the plant and conveys it to the dust-collection system. See the beginning of Chapter 9 for more information on pneumatic-conveyor systems.

Dust-Collection System

The design and configuration of the dust-collection system varies with the vendor and the specific application. Generally, a system consists of either a single large hopper-like vessel or a series of hoppers with a fan or blower affixed to the discharge manifold. Inside the vessel is an inlet manifold that directs the incoming air or gas to the dirty side of the filter media or bag. A plenum, or divider plate, separates the dirty and clean sides of the vessel.

Filter media, usually long cylindrical tubes or bags, are attached to the plenum. Depending on the design, the dust-laden air or gas may flow into the cylindrical filter bag and exit to the clean side or it may flow through the bag from its outside and exit through the tube's opening. Figure 12–1 illustrates a typical baghouse configuration.

Fabric-filter designs fall into three types, depending on the method of cleaning used: shaker cleaned, reverse-flow cleaned, and reverse-pulse cleaned.

Shaker-Cleaned Filter   The open lower ends of shaker-cleaned filter bags are fastened over openings in the tube sheet that separates the lower, dirty-gas inlet chamber from the upper, clean-gas chamber. The bags are suspended from supports that are connected to a shaking device.

The dirty gas flows upward into the filter bag and the dust collects on the inside surface. When the pressure drop rises to a predetermined upper limit due to dust accumulation, the gas flow is stopped and the shaker is operated. This process dislodges the dust, which falls into a hopper located below the tube sheet.

For continuous operation, the filter must be constructed with multiple compartments. This is necessary so that individual compartments can be sequentially taken off-line for cleaning while the other compartments continue to operate.

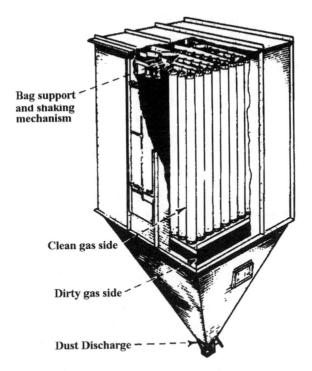

Bag support
and shaking
mechanism

Clean gas side

Dirty gas side

Dust Discharge

Figure 12–1 A typical baghouse (Perry and Green 1984).

Ordinary shaker-cleaned filters may be cleaned every 15 minutes to eight hours, depending on the service conditions. A manometer connected across the filter is used to determine the pressure drop, which indicates when the filter should be shaken. Fully automatic filters may be shaken every 2 minutes, but bag maintenance is greatly reduced if the time between shakings can be increased to 15 to 20 minutes.

The determining factor in the frequency of cleaning is the pressure drop. A differential-pressure switch can serve as the actuator in automatic cleaning applications. Cyclone precleaners sometimes are used to reduce the dust load on the filter or to remove large particles before they enter the bag.

It is essential to stop the gas flow through the filter during shaking in order for the dust to fall off. With very fine dust, it may be necessary to equalize the pressure across the cloth. In practice, this can be accomplished without interrupting continuous operation by removing from service one section at a time. With automatic filters, this operation involves closing the dirty-gas inlet dampers, shaking the filter units either pneumatically or mechanically, and reopening the dampers. In some cases, a reverse flow of clean gas through the filter is used to augment the shaker-cleaning process.

The gas entering the filter must be kept above its dewpoint to avoid water-vapor condensation on the bags, which will cause plugging. However, fabric filters have been used successfully in steam atmospheres, such as those encountered in vacuum dryers. In these applications, the housing generally is steam cased.

Reverse-Flow-Cleaned Filter   Reverse-flow-cleaned filters are similar to the shaker-cleaned design, except the shaker mechanism is eliminated. As with shaker-cleaned filters, compartments are taken off-line sequentially for cleaning. The primary use of reverse-flow cleaning is in units using fiberglass-fabric bags at temperatures above 150°C (300°F).

After the dirty-gas flow is stopped, a fan forces clean gas through the bags from the clean-gas side. The superficial velocity of the gas through the bag generally is 1.5 to 2.0 ft per minute, or about the same velocity as the dirty-gas inlet flow. This flow of clean gas partially collapses the bag and dislodges the collected dust, which falls into the hopper. Rings usually are sewn into the bags at intervals along their length to prevent complete collapse, which would obstruct the fall of the dislodged dust.

Reverse-Pulse-Cleaned Filter   In the reverse-pulse-cleaned filter, the bag forms a sleeve drawn over a cylindrical wire cage, which supports the fabric on the clean-gas side (i.e., inside) of the bag. The dust collects on the outside of the bag.

A venturi nozzle is located in the clean-gas outlet from each bag, which is used for cleaning. A jet of high-velocity air is directed through the venturi nozzle and into the bag, which induces clean gas to pass through the fabric to the dirty side. The high-velocity jet is released in a short pulse, usually about 100 milliseconds, from a compressed air line by a solenoid-controlled valve. The pulse of air and clean gas expand the bag and dislodge the collected dust. Rows of bags are cleaned in a timed sequence by programmed operation of the solenoid valves. The pressure of the pulse must be sufficient to dislodge the dust without ceasing gas flow through the baghouse.

It is common practice to clean the bags on-line without stopping the flow of dirty gas into the filter. Therefore, reverse-pulse bag filters often are built without multiple compartments. However, investigation has shown that a large fraction of the dislodged dust redeposits on neighboring bags rather than falls into the dust hopper.

As a result, there is a growing trend to clean reverse-pulse filters off-line by using bags with multiple compartments. These sections allow the outlet-gas plenum serving a particular section to be closed off from the clean-gas exhaust, thereby stopping the flow of inlet gas. On the dirty-side of the tube sheet, the isolated section is separated by partitions from neighboring sections where filtration continues. Sections of the filter are cleaned in rotation as with shaker and reverse-flow filters.

Some manufacturers design bags for use with relatively low-pressure air (i.e., 15 psi) instead of the normal 100 psi air. This allows them to eliminate the venturi tubes for clean-gas induction. Others have eliminated the separate jet nozzles located at the individual bags in favor of a single jet to pulse air into the outlet-gas plenum.

Reverse-pulse filters typically are operated at higher filtration velocities (i.e., air-to-cloth ratios) than shaker or reverse-flow designs. Filtration velocities may range from 3 to 15 ft per minute in reverse-pulse applications, depending on the dust being collected. However, the most commonly used range is 4-5 ft per minute.

The frequency of cleaning depends on the nature and concentration of the dust. Typical cleaning intervals vary from about 2 to 15 min. However, the cleaning action of the pulse is so effective that the dust layer may be completely removed from the surface of the fabric. Consequently, the fabric itself must serve as the principal filter media for a substantial part of the filtration cycle, which decreases cleaning efficiency. Because of this, woven fabrics are unsuitable for use in these devices and felt-type fabrics are used instead. With felt filters, although the bulk of the dust still is removed, an adequate level of dust collection is provided by the fabric until the dust layer reforms.

Cleaning System

As discussed in the preceding section, filter bags must be cleaned periodically to prevent excessive buildup of dust and to maintain an acceptable pressure drop across the filters. Two of the three designs discussed, reverse-flow and reverse-pulse, depend on an adequate supply of clean air or gas to provide this periodic cleaning. Two factors are critical in these systems: the clean-gas supply and the proper cleaning frequency.

Clean-Gas Supply Most applications that use the reverse-flow cleaning system use ambient air as the primary supply of clean gas. A large fan or blower draws ambient air into the clean side of the filter bags. However, unless the air is properly conditioned by inlet filters, it may contain excessive dirt loads that can affect the bag life and efficiency of the dust-collection system.

In reverse-pulse applications, most plants rely on plant-air systems as the source for the high-velocity pulses required for cleaning. In many cases, however, the plant-air system is insufficient for this purpose. Although the pulses required are short (i.e., 100 milliseconds or less), the number and frequency can deplete the supply. Therefore, care must be taken to ensure that both sufficient volume and pressure are available to achieve proper cleaning.

Cleaning Frequency Proper operation of a baghouse, regardless of design, depends on frequent cleaning of the filter media. The system is designed to operate within a specific range of pressure drops that defines clean and fully loaded filter media. The cleaning frequency must assure that the maximum recommended pressure drop is not exceeded.

This can be a real problem for baghouses that rely on automatic timers to control cleaning frequency. The use of a timing function to control cleaning frequency is not recommended unless the dust load is known to be consistent. A better approach is to use differential-pressure gauges to physically measure the pressure drop across the filter media to trigger the cleaning process based on preset limits.

Fan or Blower

All baghouse designs use some form of fan, blower, or centrifugal compressor to provide the dirty-air flow required for proper operation. In most cases, these units are installed on the clean side of the baghouse to draw the dirty air through the filter media.

Since these units provide the motive power required to transport and collect the dust-laden air, their operating condition is critical to the baghouse system. The type and size of air-moving unit varies with the baghouse type and design. Refer to the O&M manuals, as well as Chapters 8 (Fans, Blowers, and Fluidizers) and 10 (Compressors) for specific design criteria for these critical units.

Performance

The primary measure of baghouse-system performance is its ability to consistently remove dust and other particulate matter from the dirty airstream. Pressure drop and collection efficiency determine the effectiveness of these systems.

Pressure Drop

The filtration, or superficial face, velocities used in fabric filters generally are in the range of 1-10 ft per minute, depending on the type of fabric, fabric supports, and cleaning methods used. In this range, pressure drops conform to Darcy's law for streamline flow in porous media, which states that the pressure drop is directly proportional to the flow rate. The pressure drop across the fabric media and the dust layer may be expressed by

$$\Delta p = K_1 V_f + K_2 \omega V_f$$

where

> Δp = pressure drop (in. of water);
> V_f = superficial velocity through filter (ft/min);
> ω = dust loading on filter (lbm/ft^2);
> K_1 = resistance coefficient for conditioned fabric (inches of water/foot/minute);
> K_2 = resistance coefficient for dust layer (in. of water/lbm/ft/min).

Conditioned fabric maintains a relatively consistent dust-load deposit following a number of filtration and cleaning cycles. K_1 may be more than ten times the value of the resistance coefficient for the original clean fabric. If the depth of the dust layer on the fabric is greater than about $\frac{1}{16}$ in. (which corresponds to a fabric dust loading on the order of 0.1 lbm/ft^2), the pressure drop across the fabric, including the dust in the pores, usually is negligible relative to that across the dust layer alone.

In practice, K_1 and K_2 are measured directly in filtration experiments. These values can be corrected for temperature by multiplying the ratio of the gas viscosity at the desired condition to the gas viscosity at the original experimental condition.

Collection Efficiency

Under controlled conditions (e.g., in the laboratory), the inherent collection efficiency of fabric filters approaches 100 percent. In actual operation, it is determined by several variables, in particular the properties of the dust to be removed, choice of filter fabric, gas velocity, method of cleaning, and cleaning cycle. Inefficiency usually results from bags that are poorly installed, torn, or stretched from excessive dust loading and excessive pressure drop.

Installation

Most baghouse systems are provided as complete assemblies by the vendor. While the unit may require some field assembly, the vendor generally provides the structural supports, which in most cases are adequate. The only controllable installation factors that may affect performance are the foundation and connections to pneumatic conveyors and other supply systems.

Foundation

The foundation must support the weight of the baghouse. In addition, it must absorb the vibrations generated by the cleaning system. This is especially true when using the shaker-cleaning method, which can generate vibrations that can adversely affect the structural supports, foundation, and adjacent plant systems.

Connections

Efficiency and effectiveness depend on leak-free connections throughout the system. Leaks reduce the system's ability to convey dust-laden air to the baghouse. One potential source for leaks is improperly installed filter bags. Because installation varies with the type of bag and baghouse design, consult the vendor's O&M manual for specific instructions.

Operating Methods

The guidelines provided in the vendor's O&M manual should be the primary reference for proper baghouse operation. Vendor-provided information should be used because there are few common operating guidelines among the various configurations. The only general guidelines applicable to most designs are cleaning frequency and inspection and replacement of filter media.

Cleaning

As previously indicated, most bag-type filters require a precoating of particulates before they can effectively remove airborne contaminates. However, particles can completely block airflow if the filter material becomes overloaded. Therefore, the primary operating criterion is to maintain the efficiency of the filter media by controlling the cleaning frequency.

Most systems use a time-sequence to control the cleaning frequency. If the particulate load entering the baghouse is constant, this approach would be valid. However, the incoming load generally changes constantly. As a result, the straight time-sequence methodology does not provide the most efficient mode of operation.

Operators should monitor the differential-pressure gauges that measure the total pressure drop across the filter media. When the differential pressure reaches the maximum recommended level (data provided by the vendor), the operator should override any automatic timer controls and initiate the cleaning sequence.

Inspecting and Replacing Filter Media

Filter media used in dust-collection systems are prone to damage and abrasive wear. Therefore, regular inspection and replacement is needed to ensure continuous, long-term performance. Any damaged, torn, or improperly sealed bags should be removed and replaced.

A common problem associated with baghouses is improper installation of filter media. Therefore, it is important to follow the instructions provided by the vendor. If the filter bags are not properly installed and sealed, overall efficiency and effectiveness are significantly reduced.

CYCLONE SEPARATORS

A widely used type of dust-collection equipment is the cyclone separator. A "cyclone" essentially is a settling chamber in which gravitational acceleration is replaced by centrifugal acceleration. Dust-laden air or gas enters a cylindrical or conical chamber tangentially at one or more points and leaves through a central opening. The dust particles, by virtue of their inertia, tend to move toward the outside separator wall from which they are led into a receiver. Under common operating conditions, the centrifugal separating force or acceleration may range from five times gravity in very large diameter, low-resistance cyclones to 2,500 times gravity in very small, high-resistance units.

Within the range of their performance capabilities, cyclones are one of the least expensive dust-collection systems. Their major limitation is that, unless very small units are used, efficiency is low for particles smaller than five microns. Although cyclones may be used to collect particles larger than 200 microns, gravity-settling chambers or simple inertial separators usually are satisfactory and less subject to abrasion.

Configuration

The internal configuration of a cyclone separator is relatively simple. Figure 12–2 illustrates a typical cross-section of a cyclone separator, which consists of the following segments:

- Inlet area that causes the gas to flow tangentially,
- Cylindrical transition area,
- Decreasing taper that increases the air velocity as the diameter decreases,
- Central return tube to direct the dust-free air out the discharge port.

Particulate material is forced to the outside of the tapered segment and collected in a drop-leg located at the dust outlet. Most cyclones have a rotor-lock valve affixed to the bottom of the drop-leg. This is a motor-driven valve that collects the particulate material and discharges it into a disposal container.

Performance

Performance of a cyclone separator is determined by flow pattern, pressure drop, and collection efficiency.

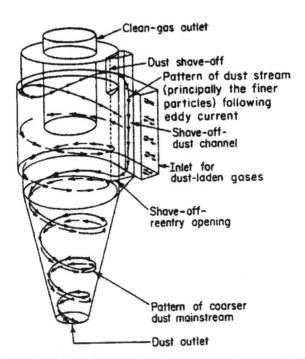

Figure 12–2 Flow pattern through a typical cyclone separator (Perry and Green 1984).

Flow Pattern

The path the gas takes in a cyclone is through a double vortex that spirals the gas downward at the outside and upward at the inside. When the gas enters the cyclone, the tangential component of its velocity, V_{ct}, increases with the decreasing radius as expressed by

$$V_{ct} \approx r^{-n}$$

In this equation, r is the cyclone radius and n is dependent on the coefficient of friction. Theoretically, in the absence of wall friction, n should equal 1.0. Actual measurements, however, indicate that n ranges from 0.5 to 0.7 over a large portion of the cyclone radius. The spiral velocity in a cyclone may reach a value several times the average inlet-gas velocity.

Pressure Drop

The pressure drop and the friction loss through a cyclone are most conveniently expressed in terms of the velocity head based on the immediate inlet area. The inlet velocity head, h_{vt}, which is expressed in inches of water, is related to the average inlet-gas velocity and density by

$$h_{vt} \approx 0.0030 \, r V_c^2$$

where

$h_{vt} =$ inlet-velocity head (in. of water);
$r \ =$ gas density (lb/ft^3);
$V_c =$ average inlet-gas velocity (ft/sec).

The cyclone friction loss, F_{cv}, is a direct measure of the static pressure and power that a fan must develop. It is related to the pressure drop by

$$F_{cv} = \Delta p_{cv} + 1 - \left(\frac{4A_c}{\pi D_e^2} \right)^2$$

where

$F_{cv} \ \ =$ friction loss (inlet-velocity heads);
$\Delta p_{cv} =$ pressure drop through the cyclone (inlet-velocity heads);
$A_c \ \ \ =$ area of the cyclone (ft^2);
$D_e \ \ \ =$ diameter of the gas exit (ft).

The friction loss through cyclones may range from 1 to 20 inlet-velocity heads, depending on its geometric proportions. For a cyclone of specific geometric proportions, F_{cv} and Δp_{cv} essentially are constant and independent of the actual cyclone size.

Collection Efficiency

Since cyclones rely on centrifugal force to separate particulates from the air or gas stream, particle mass is the dominant factor that controls efficiency. For particulates with high densities (e.g., ferrous oxides), cyclones can achieve 99 percent or better removal efficiencies, regardless of particle size. Lighter particles (e.g., tow or flake) dramatically reduce cyclone efficiency.

These devices generally are designed to meet specific pressure-drop limitations. For ordinary installations operating at approximately atmospheric pressure, fan limitations dictate a maximum allowable pressure drop corresponding to a cyclone inlet velocity in the range of 20-70 ft per second. Consequently, cyclones usually are designed for an inlet velocity of 50 ft per second.

Varying operating conditions change dust-collection efficiency by only a small amount. The primary design factor that controls collection efficiency is cyclone diameter. A small-diameter unit operating at a fixed pressure drop has a higher efficiency than a large-diameter unit. Reducing the gas-outlet duct diameter also increases the collection efficiency.

Installation

As in any other pneumatic-conveyor system, special attention must be given to the piping or ductwork used to convey the dust-laden air or gas. The inside surfaces must be smooth and free of protrusions that affect the flow pattern. All bends should be gradual and provide a laminar-flow path for the gas. See the appropriate section in Chapter 9 for specific installation information on pneumatic conveyors.

Operating Methods

Cyclones are designed for continuous operation and must be protected from plugging. In intermittent applications, the operating practices must include specific steps to purge the entire system of particulates prior to shutdown.

Pressure drop is the only factor that can be effectively controlled by an operator. Using the fan dampers, the operator can increase or decrease the cyclone's load by varying the velocity of the entering dirty air.

13

PROCESS ROLLS

Many types of process rolls are used in industrial applications. However, all share common design, installation, and operating criteria, and this chapter provides a practical review of their design and application. In general, rolls can be divided into two major classifications: working and conveying.

Working rolls change the product being processed through the production system. Included in this classification are printing rolls, which transfer a pattern to the product; corrugating rolls used to impart a profile to the product; bridle rolls, which provide torsional power to drive the product through the process; and work rolls used by the metal-processing industry to change product thickness and shape.

Conveying rolls transport the product from one point to another. This type of roll ranges from small-diameter, nondriven rolls used in simple conveyors to large-diameter, driven rolls used to transport steel, paper, and a variety of other products through continuous-process lines.

CONFIGURATION

All process rolls are composed of the following parts: body, face, neck, and bearing-support shafts. Figure 13–1 illustrates a typical process roll used in continuous-process lines.

Body

Depending on the specific application, the roll body may be constructed of a variety of materials. Typically, cast iron or steel is used, but more exotic materials, such as Monel, stainless steel, or bronze, may be used for certain applications.

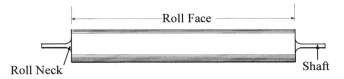

Figure 13–1 Typical process roll.

Conveying-roll bodies normally are cylindrical, but work-roll bodies may have a variety of shapes or profiles. In many of these applications, the roll body will have a specific profile, commonly referred to as a *crown*, that enhances the work performed by the roll. The profiles range from convex to concave, which determines the force transmission the roll provides.

Face

The roll face is the surface of the roll body. This is the area that performs work. Typically, the roll face is ground and polished to provide a smooth surface that does not affect the product when it is in contact with the roll.

A variety of finishing techniques are used to prepare the roll face. In work-roll applications, the face may be chrome plated, rubber coated, etched, or corrugated. The finishing method is determined by the specific application and the work to be performed. For example, coatings such as rubber commonly are used to increase friction between the roll face and the transported product. A corrugated surface is used to impart a pattern to the product (e.g., paper towels).

Neck

The neck is the transition area on both ends of the roll body that reduces the roll's diameter to that of the bearing-support shafts. The design methodology used for roll-neck construction varies with the intended function of the roll. For example, rolls used in a cold-reduction mill have a cast-steel body and neck. Because the roll must bend in normal operation, the necks are not hardened, to facilitate bending.

Neck design is critical to roll reliability, and many failures can be directly attributed to poor design. On large-diameter rolls, the reduction in diameter from the body to the final shaft size should be in steps rather than as a single reduction. Each step down should have stress-relief cuts at the transition points to prevent stress failure. Smaller-diameter rolls can be reduced in a single step, but they also must have stress relief by undercutting to prevent failure.

Bearing-Support Shafts

Many roll failures can be directly attributed to poor shaft design. In these cases, the total span from the roll body to the bearing-support point is too long for the shaft diameter. As a result, the bending moment imparted by the roll during normal operation creates an alternating compression-tension stress on the shafts. The typical failure point is where the shaft diameter changes.

Both the total bearing span from inboard to outboard bearing and the cantilevered spans from the roll body to the bearing-support point must be carefully considered when designing a process roll. The design must withstand the total forces generated in both normal and abnormal operation.

The fact that roll necks generally are relatively long and use multiple shaft-diameter reductions causes two problems. First, the long span and reduced diameter weaken the shaft, increasing the probability of excessive bending and the potential for premature failure. The second problem is the 90° corner created by the diameter reduction. This corner creates stress points that work harden when the roll is subjected to bending moments and strip tension.

A good design limits the number of shaft-diameter reductions and eliminates the 90° corners by filleting these transition points. This approach removes the stress points created by sharp corners and increases the strength of the shaft. Figure 13–2 illustrates the proper way to reduce a shaft's diameter using a stress-relief radius.

It is important to visually inspect process rolls. Poorly designed rolls and those used in improperly monitored applications are highly susceptible to premature failure. Rolls with multiple shaft reductions with or without 90° corners at these reductions warrant special attention in a predictive-maintenance program. It is important to carefully monitor strip tension, the amount of roll deflection or bending, and any other load that may be present.

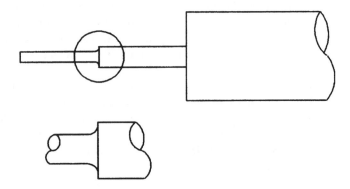

Figure 13–2 Diameter reduction of a shaft using a stress-relief radius.

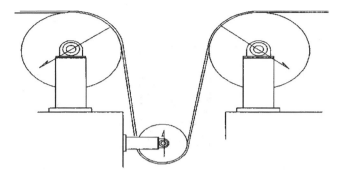

Figure 13–3 Load zones determined by wrap.

PERFORMANCE

Process rolls are subjected to variable loads induced by tension, tracking, and other process variables. Figure 13–3 illustrates the wrap of the strip as it passes over a series of rolls in a continuous process line. The load zones are indicated by arrows. In a normal operation, the force or load induced by the strip is uniformly applied across the roll's entire face or body, as illustrated in Figure 13–4.

The width of the product or belt in contact with the roll has a direct effect on loading and how the load is transmitted to the roll and its bearing-support structures. Figure 13–5 illustrates a narrow strip that is tracking properly. Note that, with a narrow strip, the load is concentrated more in the center of the roll and not uniformly transferred across the entire roll face. This tends to bend the roll, with the degree of deflection depending on three factors: roll diameter, roll construction, and strip tension.

Figure 13–4 Strip uniformly loads roll.

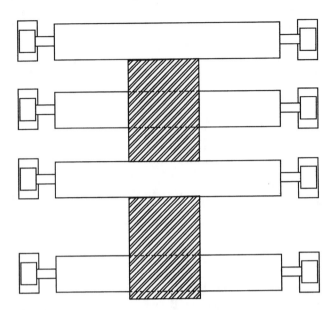

Figure 13–5 Narrow strip concentrates load toward the center of the roll.

INSTALLATION

The proper installation of process rolls is critical. As with all other machines, alignment and proper bolting techniques are extremely important. Misalignment can cause poor quality, reduced capacity, and premature failure.

Single Rolls

With the exception of steering rolls, all single rolls in a continuous-process line or conveyor system must be perpendicular to the centerline of the belt, strip, or conveyed product (i.e., passline) and have the same elevation on both the operator and drive sides. Any misalignment, either horizontal or vertical, influences the tracking of the belt, strip, and conveyed product.

Figure 13–6 illustrates a roll that has uneven elevation, or is vertically misaligned. With this type of misalignment, the strip will have greater tension on the side of the roll with the higher elevation, forcing it to move toward the lower end. In effect, the roll becomes a steering roll, forcing the strip, belt, or product to one side of the centerline of travel.

Paired Rolls

Rolls designed to work in pairs (e.g., scrubber, corrugator, or printing rolls) must be perpendicular and level to the passline. In addition, they must be parallel to each

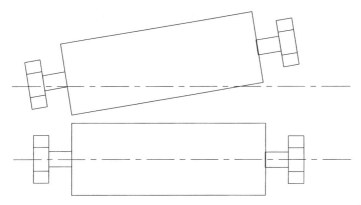

Figure 13–6 A roll with uneven elevation becomes a steering roll.

other. Figure 13–7 illustrates a paired set of scrubber rolls in which the strip is captured between the two rolls and the counter-rotating brush roll cleans the strip surface.

The design of some roll pairs makes it difficult to keep them parallel. Many have a single pivot point to fix one end of the roll and a pneumatic cylinder to index the opposite end. Others use two cylinders, one attached to each end of the roll. In this design, the two cylinders are not mechanically linked and the rolls do not maintain their parallel relationship.

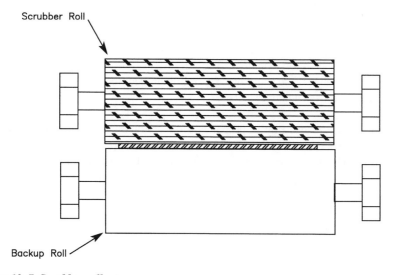

Scrubber Roll

Backup Roll

Figure 13–7 Scrubber roll set.

Figure 13–8 Result of misalignment (i.e., nonparallel operation).

Nonparallel operation of paired rolls reduces their life. For example, the scrubber/ backup roll set should provide extended service life. However, the brush rolls some-times have a life of only a few weeks. The brush roll wear profile shown in Figure 13–8 is clear evidence of nonparallel operation. After a very short service life, the brush rolls become conical in shape, much like a bottle brush. This wear pattern is visual conformation that the brush roll and its mating rubber-coated backup roll are not parallel.

Poor alignment has another negative affect on rolls. It changes the load zone so that one of the rolls must absorb more of the load than the other. In conveying applica-tions, this may not seriously affect roll life or cause catastrophic failure. However, in applications where the roll must support extreme loads or impart large forces, like corrugators or rolling mills, this shift in the load zone can result in catastrophic roll failure.

OPERATING METHODS

Abnormal induced loads is the most serious operator-controlled variable that affects roll performance. Operating methods should ensure that proper tension, product tracking, and torsional loads are maintained at all times.

In particular, operators should look for torsional-load variations caused by speed and load changes. In most cases, roll damage occurs when they are subjected to momen-tary radical changes in torsional load. These transients normally occur during startup, process-speed changes, and shutdown. Operating procedures should be developed and followed to minimize such transients.

Abnormal loading caused by improper tension or tracking of the product, belt, or other media carried by the rolls also will reduce the performance and useful life of process rolls. The load induced by the carried media should be equally and uniformly distributed across the entire roll face. If the load is concentrated off-center, it will cause premature wear and possible failure of the bearing, roll neck, and shaft. Operat-ing procedures should provide positive methods that monitor and correct abnormal tension or tracking.

14

GEARBOXES/REDUCERS

A gear is a disk or wheel with teeth around its periphery, either on the inside edge (i.e., internal gear) or on the outside edge (i.e., external gear). A gear provides a positive means of power transmission, which is effected by the teeth on one gear meshing with the teeth on another gear or rack (i.e., straight-line gear).

Gear drives are packaged units used for a wide range of power-transmission applications. They are used to transmit power to a driven piece of machinery and to change or modify the power that is transmitted. Modifications include reducing speed and increasing output torque, increasing speed, changing the direction of shaft rotation, or changing the angle of shaft operation.

CONFIGURATION

Several different types of gears are used in industry. Many are complex in design and manufacture and several have evolved directly from the spur gear, referred to as the *basic* gear. Types of gears are spur, helical, bevel, and worm. Table 14–1 summarizes the characteristics of each gear type.

Spur

The spur gear is the least expensive of all gears to manufacture and the one most commonly used. It can be manufactured to close tolerances and is used to connect parallel shafts that rotate in opposite directions. The spur gear gives excellent results at moderate peripheral speeds, and the tooth load produces no axial thrust. Because contact is simultaneous across the entire width of the meshing teeth, it tends to be noisy at high speeds. However, noise and wear can be minimized with proper lubrication.

Table 14–1 Gear Characteristics Overview

Gear Type	Characteristics	
	Attributes/Positives	**Negatives**
Spur, external	Connects parallel shafts that rotate in opposite directions, inexpensive to manufacture to close tolerances, moderate peripheral speeds, no axial thrust, high mechanical efficiency	Noisy at high speeds
Spur, internal	Compact drive mechanism for parallel shafts rotating in same direction	
Helical, external	Connects parallel and nonparallel shafts; superior to spur gears in load-carrying capacity, quietness, and smoothness; high efficiency	Higher friction than spur gears, high end thrust
Helical, double (also referred to as herringbone)	Connects parallel shafts, overcomes high-end thrust present in single-helical gears, compact, quiet and smooth operation at higher speeds (1,000 to 12,000 fpm or higher), high efficiencies	
Helical, cross	Light loads with low power transmission demands	Narrow range of applications, requires extensive lubrication
Bevel	Connects angular or intersecting shafts	Gears overhang supporting shafts resulting in shaft deflection and gear misalignment
Bevel, straight	Peripheral speeds up to 1,000 fpm in applications where quietness and maximum smoothness not important, high efficiency	Thrust load causes gear pair to separate
Bevel, zerol	Same ratings as straight bevel gears and uses same mountings, permits slight errors in assembly, permits some displacement due to deflection under load, highly accurate, hardened due to grinding	Limited to speeds less than 1,000 fpm due to noise
Bevel, spiral	Smoother and quieter than straight bevel gears at speeds greater than 1,000 fpm or 1,000 rpm, evenly distributed tooth loads, carry more load without surface fatigue, high efficiency, reduces size of installation for large reduction ratios, speed-reducing and speed-increasing drive	High tooth pressure, thrust loading depends on rotation and spiral angle

Table 14–1 Gear Characteristics Overview (continued)

Gear Type	Characteristics	
	Attributes/Positives	**Negatives**
Bevel, miter	Same number of teeth in both gears, operate on shafts at 90°	
Bevel, hypoid	Connects nonintersecting shafts, high pinion strength, allows the use of compact straddle mounting on the gear and pinion, recommended when maximum smoothness required, compact system even with large reduction ratios, speed-reducing and speed-increasing drive	Lower efficiency, difficult to lubricate due to high tooth-contact pressures, materials of construction (steel) require use of extreme-pressure lubricants
Planetary or epicyclic	Compact transmission with driving and driven shafts in line, large speed reduction when required	
Worm, cylindrical	Provide high-ratio speed reduction over wide range of speed ratios (60:1 and higher from a single reduction, can go as high as 500:1), quiet transmission of power between shafts at 90°, reversible unit available, low wear, can be self-locking	Lower efficiency; heat removal difficult, which restricts use to low-speed applications
Worm, double-enveloping	Increased load capacity	Lower efficiencies

Source: Integrated Systems, Inc.

There are three main classes of spur gears: external tooth, internal tooth, and rack-and-pinion. The external tooth variety shown in Figure 14–1 is the most common. Figure 14–2 illustrates an internal gear, and Figure 14–3 shows a rack or straight-line spur gear.

The spur gear is cylindrical and has straight teeth cut parallel to its rotational axis. The tooth size of spur gears is established by the diametrical pitch. Spur-gear design accommodates mostly rolling, rather than sliding, contact of the tooth surfaces and tooth contact occurs along a line parallel to the axis. Such rolling contact produces less heat and yields high mechanical efficiency, often up to 99 percent.

An internal spur gear, in combination with a standard spur-gear pinion, provides a compact drive mechanism for transmitting motion between parallel shafts that rotate

Figure 14–1 Example of a spur gear (Neale 1993).

in the same direction. The internal gear is a wheel that has teeth cut on the inside of its rim and the pinion is housed inside the wheel. The driving and driven members rotate in the same direction at relative speeds inversely proportional to the number of teeth.

Helical

Helical gears, shown in Figure 14–4, are formed by cutters that produce an angle that allows several teeth to mesh simultaneously. Helical gears are superior to spur gears in their load-carrying capacity, quietness, and smoothness of operation, which results

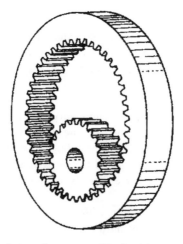

Figure 14–2 Example of an internal spur gear (Neale 1993).

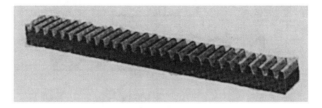

Figure 14–3 Rack or straight-line gear (Neale 1993).

from the sliding contact of the meshing teeth. A disadvantage, however, is the higher friction and wear that accompanies this sliding action.

Single helical gears are manufactured with the same equipment as spur gears, but the teeth are cut at an angle to the axis of the gear and follow a spiral path. The angle at which the gear teeth are cut is called the *helix angle*, which is illustrated in Figure 14–5. This angle causes the position of tooth contact with the mating gear to vary at each section. Figure 14–6 shows the parts of a helical gear.

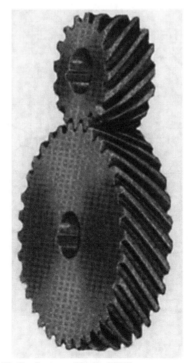

Figure 14–4 Typical set of helical gears (Neale 1993).

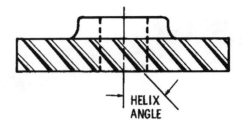

Figure 14–5 Illustrating the angle at which the teeth are cut (Neale 1993).

It is very important to note that the helix angle may be on either side of the gear's center line. Or, if compared to the helix angle of a thread, it may be either a "right-hand" or "left-hand" helix. Figure 14–7 illustrates a helical gear as viewed from opposite sides. A pair of helical gears must have the same pitch and helix angle but be of opposite hand (one right hand and one left hand).

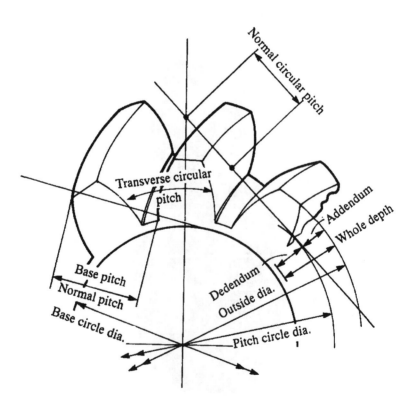

Figure 14–6 Helical gear and its parts (95/96 Product Guide).

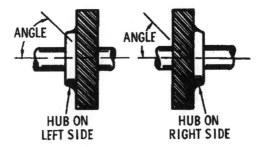

Figure 14–7 The helix angle of the teeth must be the same no matter from which side the gear is viewed (Neale 1993).

Herringbone

The double-helical gear, also referred to as the *herringbone gear* (Figure 14–8), is used for transmitting power between parallel shafts. It was developed to overcome the disadvantage of the high-end thrust present with single-helical gears.

The herringbone gear consists of two sets of gear teeth on the same gear, one right hand and one left hand. Having both hands of gear teeth causes the thrust of one set to cancel out the thrust of the other. Therefore, another advantage of this gear type is quiet, smooth operation at higher speeds.

Bevel

Bevel gears are used most frequently for 90° drives, but other angles can be accommodated. The most typical application is driving a vertical pump with a horizontal driver.

Figure 14–8 Herringbone gear (Neale 1993).

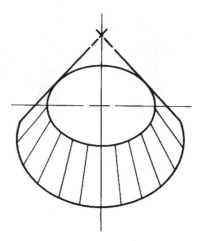

Figure 14–9 Basic cone shape of bevel gears (Neale 1993).

Two major differences between bevel gears and spur gears are their shape and the relation of the shafts on which they are mounted. A bevel gear is conical in shape, while a spur gear is essentially cylindrical. The diagram in Figure 14–9 illustrates the bevel gear's basic shape. Bevel gears transmit motion between angular or intersecting shafts, while spur gears transmit motion between parallel shafts.

Figure 14–10 shows a typical pair of bevel gears. As with other gears, the term *pinion and gear* refers to the members with the smaller and larger numbers of teeth in the

Figure 14–10 Typical set of bevel gears (Neale 1993).

pair, respectively. Special bevel gears can be manufactured to operate at any desired shaft angle, as shown in Figure 14–11.

As with spur gears, the tooth size of bevel gears is established by the diametrical pitch. Because the tooth size varies along its length, measurements must be taken at a specific point. Note that, because each gear in a bevel-gear set must have the same pressure angle, tooth length, and diametrical pitch, they are manufactured and distributed only as mated pairs. Like spur gears, bevel gears are available in pressure angles of 14.5° and 20°.

Because there generally is no room to support bevel gears at both ends due to the intersecting shafts, one or both gears overhang their supporting shafts. This, referred to as an *overhung load*, may result in shaft deflection and gear misalignment, causing poor tooth contact and accelerated wear.

Straight or Plain

Straight-bevel gears, also known as *plain bevels*, are the most commonly used and simplest type of bevel gear (Figure 14–12). They have teeth cut straight across the face of the gear. These gears are recommended for peripheral speeds up to 1,000 ft per minute in cases where quietness and maximum smoothness are not crucial. This gear type produces thrust loads in a direction that tends to cause the pair to separate.

Zerol

Zerol-bevel gears are similar to straight-bevel gears, carry the same ratings, and can be used in the same mountings. These gears, which should be considered spiral-bevel gears having a spiral angle of zero, have curved teeth that lie in the same general

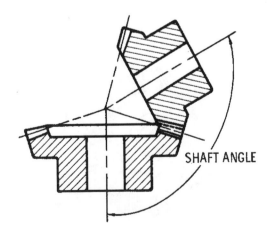

Figure 14–11 Shaft angle, which can be at any degree (Neale 1993).

Figure 14–12 Straight or plain bevel gear (Neale 1993).

direction as straight-bevel gears. This type of gear permits slight errors in assembly and some displacement due to deflection under load. Zerol gears should be used at speeds less than 1,000 ft per minute because of excessive noise at higher speeds.

Spiral

Spiral-bevel gears (Figure 14–13) have curved oblique teeth that contact each other gradually and smoothly from one end of the tooth to the other, meshing with a rolling contact similar to helical gears. Spiral-bevel gears are smoother and quieter in operation than straight-bevel gears, primarily due to a design that incorporates two or more contacting teeth. Their design, however, results in high tooth pressure.

This type of gear is beginning to supersede straight-bevel gears in many applications. They have the advantage of ensuring evenly distributed tooth loads and carry more load without surface fatigue. Thrust loading depends on the direction of rotation and whether the spiral angle of the teeth is positive or negative.

Figure 14–13 Spiral bevel gear (Neale 1993).

Figure 14–14 Miter gear shaft angle (Neale 1993).

Miter

Miter gears are bevel gears with the same number of teeth in both gears, operating on shafts at right angles, or 90°, as shown in Figure 14–14. Their primary use is to change direction in a mechanical drive assembly. Since both the pinion and gear have the same number of teeth, no mechanical advantage is generated by this type of gear.

Hypoid

Hypoid-bevel gears are a cross between a spiral-bevel gear and a worm gear (Figure 14–15). The axes of a pair of hypoid-bevel gears are nonintersecting and the distance between the axes is referred to as the *offset*. This configuration allows both shafts to be supported at both ends and provides high strength and rigidity.

Although stronger and more rigid than most other types of gears, they are less efficient and extremely difficult to lubricate because of high tooth-contact pressures. Further

Figure 14–15 Hypoid bevel gear (Neale 1993).

increasing the demands on the lubricant is the material of construction, as both the driven and driving gears are made of steel. This requires the use of special extreme-pressure lubricants that have both oiliness and antiweld properties that can withstand the high contact pressures and rubbing speeds.

Despite its demand for special lubrication, this gear type is in widespread use in industrial and automotive applications. It is used extensively in rear axles of automobiles having rear-wheel drives and increasingly is being used in industrial machinery.

Worm

The worm and gear, which are illustrated in Figure 14–16, are used to transmit motion and power when a high-ratio speed reduction is required. They accommodate a wide range of speed ratios (60:1 and higher can be obtained from a single reduction and can go as high as 500:1). In most worm-gear sets, the worm is the driver and the gear the driven member. They provide a steady, quiet transmission of power between shafts at right angles and can be self-locking. Thus, torque on the gear will not cause the worm to rotate.

The contact surface of the screw on the worm slides along the gear teeth. Because of the high level of rubbing between the worm and wheel teeth, however, slightly less efficiency is obtained than with precision spur gears. Note that large helix angles on the gear teeth produce higher efficiencies. Another problem with this gear type is heat removal, a limitation that restricts their use to low-speed applications.

Figure 14–16 Worm gear (Nelson 1986).

A major advantage of the worm gear is low wear, due mostly to a full-fluid lubricant film. In addition, friction can be further reduced through the use of metals having low coefficients of friction. For example, the wheel typically is made of bronze and the worm of a highly finished hardened steel.

Most worms are cylindrical in shape with a uniform pitch diameter. However, a variable pitch diameter is used in the double-enveloping worm. This configuration is used when increased load capacity is required.

PERFORMANCE

With few exceptions, gears are one-directional power transmission devices. Unless a special, bidirectional gear set is specified, gears have a specific direction of rotation and will not provide smooth, trouble-free power transmission when the direction is reversed. The reason for this one-directional limitation is that gear manufacturers do not finish the nonpower side of the tooth profile. This is primarily a cost-savings issue and should not affect gear operation.

The primary performance criteria for gear sets include efficiency, brake horsepower, speed transients, startup, backlash, and ratios.

Efficiency

Gear efficiency varies with the type of gear used and the specific application. Table 14–2 provides a comparison of the approximate efficiency range of various gear types. The table assumes normal operation, where torsional loads are within the gear set's designed horsepower range. It also assumes that startup and speed change torques are acceptable.

Table 14–2 Gear Efficiencies

Gear Type	Efficiency Range (%)
Bevel gear, hypoid	90–98
Bevel gear, miter	Not available
Bevel gear, spiral	97–99
Bevel gear, straight	97–99
Bevel gear, zerol	Not available
Helical gear, external	97–99
Helical gear–double, external (herringbone)	97–99
Spur gear, external	97–99
Worm, cylindrical	50–99
Worm, double-enveloping	50–98

Source: Adapted by Integrated Systems, Inc., from "Gears and Gear Drives," *1996 Power Transmission Design* (Penton Publishing Inc., 1996), pp. A199–A211.

Brake Horsepower

All gear sets have a recommended and maximum horsepower rating. The rating varies with the type of gear set but must be carefully considered when evaluating a gearbox problem. The maximum installed motor horsepower should never exceed the maximum recommended horsepower of the gearbox. This is especially true of worm gear sets. The soft material used for these gears is damaged easily when excess torsional load is applied.

The procurement specifications or the vendor's engineering catalog will provide all the recommended horsepower ratings needed for an analysis. These recommendations assume normal operation and must be adjusted for the actual operating conditions in a specific application.

Speed Transients

Applications that require frequent speed changes can have a severe, negative impact on gearbox reliability. The change in torsional load caused by acceleration and deceleration of a gearbox may exceed its maximum allowable horsepower rating. This problem can be minimized by decreasing the ramp speed and amount of braking applied to the gear set. The vendor's O&M manual or technical specifications should provide detailed recommendations that define the limits to use in speed-change applications.

Startup

Start-stop operation of a gearbox can accelerate both gear and bearing wear and may cause reliability problems. In applications like the bottom discharge of storage silos, where a gear set drives a chain or screw conveyor system and startup torque is excessive, care must be taken to prevent overloading the gear set.

Backlash

Gear backlash is the play between teeth measured at the pitch circle. It is the distance between the involutes of the mating gear teeth as illustrated in Figure 14–17.

Backlash is necessary to provide the running clearance needed to prevent binding of the mating gears, which can result in heat generation, noise, abnormal wear, overload, and/or failure of the drive. In addition to the need to prevent binding, some backlash occurs in gear systems because of the dimensional tolerances needed for cost-effective manufacturing.

During the gear-manufacturing process, backlash is achieved by cutting each gear tooth thinner by an amount equal to one half the backlash dimension required for the application. When two gears made in this manner are run together (i.e., mate), their allowances combine to provide the full amount of backlash.

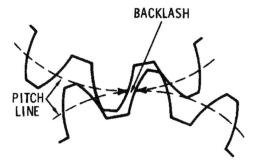

Figure 14–17 Backlash (Neale 1993).

The increase in backlash that results from tooth wear does not adversely affect operation with nonreversing drives or drives with a continuous load in one direction. However, for reversing drives and drives where timing is critical, excessive backlash that results from wear usually cannot be tolerated.

Ratios

Gears are defined and specified using the gear-tooth ratio, contact ratio, and hunting ratio. The gear-tooth ratio is the ratio of the larger to the smaller number of teeth in a pair of gears. The contact ratio is a measure of overlapping tooth action, which is necessary to assure smooth, continuous action. For example, as one pair of teeth passes out of action, a succeeding pair of teeth already must have started action. The hunting ratio is the ratio of the number of gear and pinion teeth. It is a means of ensuring that every tooth in the pinion contacts every tooth in the gear before it contacts any gear tooth a second time.

INSTALLATION

Installation guidelines provided in the vendor's O&M manual should be followed for proper installation of the gearbox housing and alignment to its mating machine-train components.

Gearboxes must be installed on a rigid base that prevents flexing of its housing and the input and output shafts. Both the input and output shaft must be properly aligned, within 0.002 in., to their respective mating shafts. Both shafts should be free of any induced axial forces that may be generated by the driver or driven units.

Internal alignment also is important. Internal alignment and clearance of new gearboxes should be within the vendor's acceptable limits, but there is no guarantee that this will be true. All internal clearance (e.g., backlash and center-to-center distances) and the parallel relationship of the pinion and gear shafts should be verified for any gearbox that is being investigated.

OPERATING METHODS

Two primary operating parameters govern effective operation of gear sets or gearboxes: maximum torsional power rating and transitional torsional requirements.

Each gear set has a specific maximum horsepower rating. This is the maximum torsional power that the gear set can generate without excessive wear or gear damage. Operating procedures should ensure that the maximum horsepower is not exceeded throughout the entire operating envelope. If the gear set was properly designed for the application, its maximum horsepower rating should be suitable for steady-state operation at any point within the design operating envelope. As a result, it should be able to provide sufficient torsional power at any set point within the envelope.

Two factors may cause overload on a gear set: excessive load or speed transients. Many processes are subjected to radical changes in the process or production loads. These changes can have a serious effect on gear-set performance and reliability.

Operating procedures should establish boundaries that limit the maximum load variations that can be used in normal operation. These limits should be well within the acceptable load rating of the gear set.

The second factor, speed transients, is a leading cause of gear-reliability problems. The momentary change in torsional load created by rapid changes in speed can have a dramatic, negative impact on gear sets. These transients often exceed the maximum horsepower rating of the gears and may result in failure. Operating procedures should ensure that torsional power requirements during startup, process-speed changes, and shutdown do not exceed the recommended horsepower rating of the gear set.

15

STEAM TRAPS

Steam-supply systems commonly are used in industrial facilities as a general heat source as well as a heat source in pipe and vessel tracing lines used to prevent freeze-up in nonflow situations. Inherent with the use of steam is the problem of condensation and the accumulation of noncondensable gases in the system.

Steam traps must be used to automatically purge condensable and noncondensable gases, such as air, from the steam system. However, a steam trap should never discharge live steam. Such discharges are dangerous as well as costly.

CONFIGURATION

Five major types of steam traps commonly are used in industrial applications: inverted bucket, float and thermostatic, thermodynamic, bimetallic, and thermostatic. Each type of steam trap uses a different method to determine when and how to purge the system. As a result, each has a different configuration.

Inverted Bucket

The inverted-bucket trap, shown in Figure 15–1, is a mechanically actuated steam trap that uses an upside down, or inverted, bucket as a float. The bucket is connected to the outlet valve through a mechanical link. The bucket sinks when condensate fills the steam trap, which opens the outlet valve and drains the bucket. It floats when steam enters the trap and closes the valve.

As a group, inverted-bucket traps can handle a wide range of steam pressures and condensate capacities. They are an economical solution for low- to medium-pressure and medium-capacity applications, such as plant heating and light processes. When used

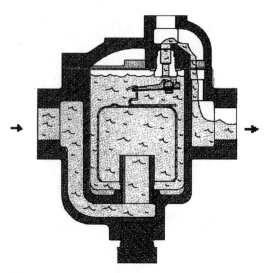

Figure 15–1 Inverted-bucket trap.

for higher-pressure and higher-capacity applications, these traps become large, expensive, and difficult to handle.

Each specific steam trap has a finite, relatively narrow range that it can handle effectively. For example, an inverted-bucket trap designed for up to 15-psi service will fail to operate at pressures above that value. An inverted-bucket trap designed for 125-psi service will operate at lower pressure, but its capacity is so diminished that it may back up the system with unvented condensate. Therefore, it is critical to select a steam trap designed to handle the application's pressure, capacity, and size requirements.

Float and Thermostatic

The float-and-thermostatic trap shown in Figure 15–2 is a hybrid. A float similar to that found in a toilet tank operates the valve. As condensate collects in the trap, it lifts the float and opens the discharge or purge valve. This design opens the discharge only as much as necessary. Once the built-in thermostatic element purges noncondensable gases, it closes tightly when steam enters the trap. The advantage of this type of trap is that it drains condensate continuously.

Like the inverted-bucket trap, float-and-thermostatic traps as a group handle a wide range of steam pressures and condensate loads. However, each individual trap has a very narrow range of pressures and capacities. This makes it critical to select a trap that can handle the specific pressure, capacity, and size requirements of the system.

The key advantage of float-and-thermostatic traps is their ability for quick steam-system startup because they continuously purge the system of air and other noncondens-

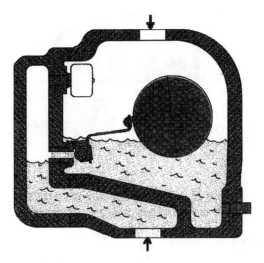

Figure 15–2 Float-and-thermostatic trap.

able gases. One disadvantage is the sensitivity of the float ball to damage by hydraulic hammer.

Float-and-thermostatic traps are an economical solution for lighter condensate loads and lower pressures. However, when the pressure and capacity requirements increase, the physical size of the unit increases and its cost rises. It also becomes more difficult to handle.

Thermodynamic or Disk Type

Thermodynamic, or disk-type, steam traps use a flat disk that moves between a cap and seat (see Figure 15–3). On startup, condensate flow raises the disk and opens the discharge port. Steam or very hot condensate entering the trap seats the disk. It remains seated, closing the discharge port, as long as pressure is maintained above it. Heat radiates out through the cap, thus diminishing the pressure over the disk, opening the trap to discharge condensate.

Wear and dirt are particular problems with a disk-type trap. Because of the large, flat seating surfaces, any particulate contamination, such as dirt or sand, will lodge between the disk and the valve seat. This prevents the valve from sealing and permits live steam to flow through the discharge port. If pressure is not maintained above the disk, the trap will cycle frequently. This wastes steam and can cause the device to fail prematurely.

The key advantage of these traps is that one trap can handle a complete range of pressures. In addition, they are relatively compact for the amount of condensate

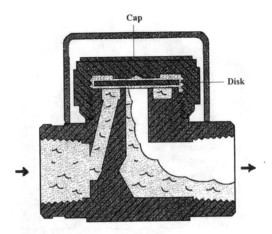

Figure 15–3 Thermodynamic steam trap.

they discharge. The chief disadvantage is difficulty in handling air and other noncondensable gases.

Bimetallic

A bimetallic steam trap, shown in Figure 15–4, operates on the same principle as a residential-heating thermostat. A bimetallic strip, or wafer, connected to a valve disk bends or distorts when subjected to a change in temperature. When properly cali-

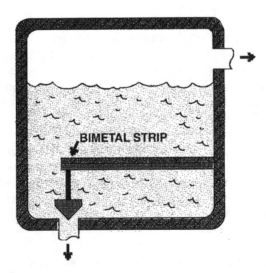

Figure 15–4 Bimetal trap.

brated, the disk closes tightly against a seat when steam is present and opens when condensate, air, and other gases are present.

Two key advantages of bimetallic traps are their compact size relative to their condensate load-handling capabilities and immunity to hydraulic-hammer damage.

Their biggest disadvantage is the need for constant adjustment or calibration, which usually is done at the factory for the intended steam operating pressure. If the trap is used at a lower pressure, it may discharge live steam. If used at a higher pressure, condensate may back up into the steam system.

Thermostatic or Thermal Element

Thermostatic, or thermal-element, traps are thermally actuated using an assembly constructed of high-strength, corrosion-resistant stainless steel plates seam-welded together. Figure 15–5 shows this type of trap.

On startup, the thermal element is positioned to open the valve and purge condensate, air, and other gases. As the system warms up, heat generates pressure in the thermal element, causing it to expand and throttle the flow of hot condensate through the discharge valve. The steam that follows the hot condensate into the trap expands the thermal element with great force, which causes the trap to close. Condensate that enters the trap during system operation cools the element. As the thermal element cools, it lifts the valve off the seat and allows the condensate to discharge quickly.

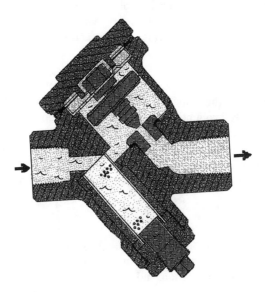

Figure 15–5 Thermostatic trap.

Thermal elements can be designed to operate at any steam temperature. In steam-tracing applications, it may be desirable to allow controlled amounts of condensate to back up in the lines in order to extract more heat from the condensate. In other applications, any hint of condensate in the system is undesirable. The thermostatic trap can handle either condition, but the thermal element must be properly selected to accommodate the specific temperature range of the application.

Thermostatic traps are compact, and a given trap operates over a wide range of pressures and capacities. However, they are not recommended for condensate loads over 15,000 lb per hour.

PERFORMANCE

When properly selected, installed, and maintained, steam traps are relatively trouble free and highly efficient. The critical factors that affect efficiency include capacity and pressure ratings, steam quality, mechanical damage, and calibration.

Capacity Rating

Each type and size of steam trap has a specified capacity for the amount of condensate and noncompressible gas that it can handle. Care must be taken to ensure that the proper steam trap is selected to meet the application's capacity needs.

Pressure Rating

As discussed previously, each type of steam trap has a range of steam pressures that it can handle effectively. Therefore, each application must be carefully evaluated to determine the normal and maximum pressures that will be generated by the steam system. Traps must be selected for a worst-case scenario.

Steam Quality

Steam quality determines the amount of condensate to be handled by the steam trap. In addition to an increased volume of condensate, poor steam quality may increase the amount of particulate matter present in the condensate. High concentrations of solids directly affect the performance of steam traps. If particulate matter is trapped between the purge valve and its seat, the steam trap may not properly shut off the discharge port. This will result in live steam being continuously exhausted through the trap.

Mechanical Damage

Inverted-bucket and float-type steam traps are highly susceptible to mechanical damage. If the level arms or mechanical links are damaged or distorted, the trap cannot operate properly. Regular inspection and maintenance of these types of traps are essential.

Calibration

Steam traps, such as the bimetallic type, must be periodically recalibrated to ensure proper operation. All steam traps should be adjusted on a regular schedule.

INSTALLATION

Installation of steam traps is relatively straightforward. As long as they are properly sized, the only installation imperative is that they be plumb. If the trap is tilted or cocked, the bucket, float, or thermal valve will not operate properly. In addition, a nonplumb installation may prevent the condensate chamber from fully discharging accumulated liquids.

OPERATING METHODS

Steam traps are designed for a relatively constant volume, pressure, and condensate load. Operating practices should attempt to maintain these parameters as much as possible. Actual operating practices are determined by the process system, rather than the trap selected for a specific system.

The operator should periodically inspect them to ensure proper operation. Special attention should be given to the drain line to ensure that the trap is properly seated when not in the bleed or vent position.

A common failure mode of steam traps is failure of the sealing device (i.e., plunger, disk, or valve) to return to a leak-tight seat when in its normal operating mode. Leakage during normal operation may lead to abnormal operating costs or degradation of the process system. A single $3/4$-in. steam trap that fails to seat properly can increase operating costs by $40,000 to $50,000 per year. Traps that fail to seat properly or are constantly in an unloading position should be repaired or replaced as quickly as possible. Regular inspection and adjustment programs should be included in the standard operating procedures.

16

INVERTERS

Inverters control the output speed of alternating current (AC) motors. While the basic function of all inverters is the same, the approach varies with the type of inverter.

CONFIGURATION

Two basic types of inverters commonly are used in industrial applications: volts/hertz and vector control.

Volts/Hertz Control

Traditionally, a volts/hertz speed-control device uses a volts-per-hertz (V/Hz) controller, which uses a mechanical-reference command taken from a shaft encoder, or resolver, to vary the voltage and frequency applied to the motor. By maintaining a constant V/Hz ratio, the inverter drive controls the speed of the connected motor. Figure 16–1 shows how this type of controller limits current frequency to the motor.

Inside the drive shown in Figure 16–1, a current-limit block monitors motor current and alters the frequency command when the motor current exceeds a predetermined value. Early V/Hz inverters were sensitive to variations in applied load and could not maintain consistent speed control in applications subjected to frequency-load variations.

The introduction of slip compensation, a feature added to later V/Hz models, altered the frequency reference to keep the actual motor speed close to the desired speed during load changes. The slip-compensation module compares the deviation between actual and no-load speed of the motor and enters a correction factor to the inverter drive. This factor compensates for the variation in speed, or slip, caused by load changes.

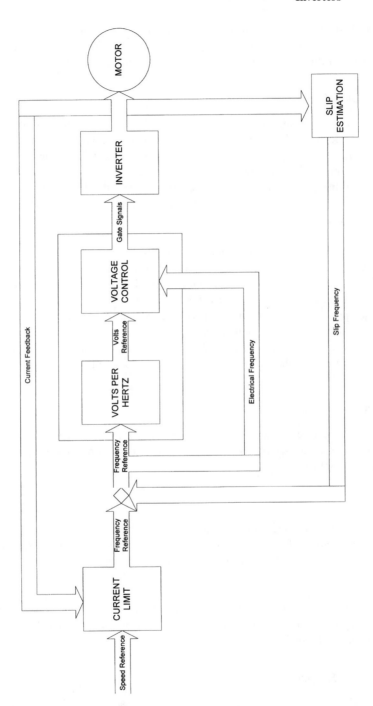

Figure 16–1 Volts/Hertz inverter limits current frequency to the motor.

Volts-per-hertz technology works well in general-purpose, moderate-speed applications. However, it is unsuitable for applications that require high dynamic response and torque control or when the motor is running at very low speeds.

Vector Control

Vector-control technology was developed to provide the ability to accurately control the output speed of alternating-current motors in both high-torque and low-speed applications. Alternating-current vector controls refer to the drive's ability to control the vector sum of flux and torque in the controlled motor, which provides precise speed and torque performance. These capabilities enable the drive to maintain tension when a machine stops or to quickly return to full speed when a heavy load variation is imposed on the driven machine.

Three basic types of vector drives commonly are used in these applications: flux-vector, voltage-vector, and stator-flux-vector controls. All these control technologies may retain the volts-per-hertz core logic, but add other control blocks to improve drive performance. These additional control blocks include a current resolver that estimates the flux- and torque-producing currents in the motor and enters a correction factor to the V/Hz primary-control logic. Where more accurate speed control is required, a current regulator may be used to replace the standard V/Hz current-limit block. In this configuration, shown in Figure 16–2, the output of the current regulator is still a frequency reference.

PERFORMANCE

Inverter performance is measured by the response characteristics of the motor. In most cases, these characteristics include torque response, impact-load response, and acceleration control.

Torque Response

Figure 16–3 illustrates the normal torque-response characteristics of a V/Hz inverter. Note that the ability of the drive to maintain high torque output at low speeds drops off significantly below 3 Hz. For this reason, the operating range of a V/Hz inverter is usually less than 20 to 1 (i.e., 20:1).

A flux-vector control improves the drive's dynamic response and may be able to control both the output torque and speed. Figure 16–4 provides a typical torque-speed response curve of a flux-vector inverter.

Impact-Load Response

Inverter drives must compensate for variations in load. Figure 16–5 compares the impact-load response of a standard V/Hz and a sensorless flux-vector-type inverter. In

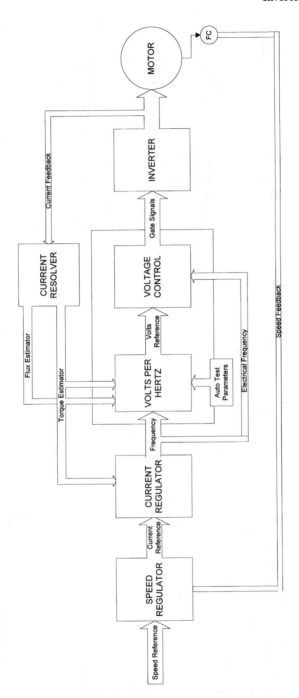

Figure 16–2 Flux-vector control with feedback loop.

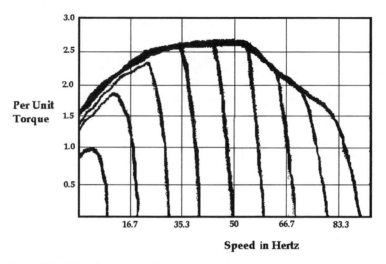

Figure 16–3 V/Hz drive cannot apply full torque as speed approaches zero.

most cases, the flux-vector inverter will have better response characteristics than the V/Hz inverter. The feedback logic used in flux-vector inverters provide a more positive means for both detection and compensation for load variations.

Acceleration Control

Inverters must provide positive speed and torque control over the full operating range of the controlled motor and system. Inverter speed-control characteristics, especially

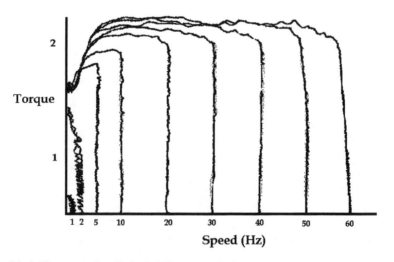

Figure 16–4 Flux vector has limited ability to supply full torque at low speeds.

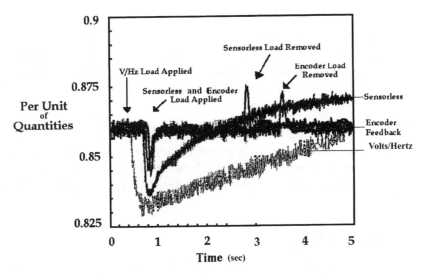

Figure 16–5 Impact load-response comparison for V/Hz and sensorless, flux-vector-type inverters.

during rapid acceleration, are important performance measurements. Figure 16–6 shows the response characteristics of a flux-vector-type inverter, which can supply three times full torque in less than 1 sec. The ability to maintain full torque, as well as the ramp or acceleration rate of the motor, is essential in critical applications.

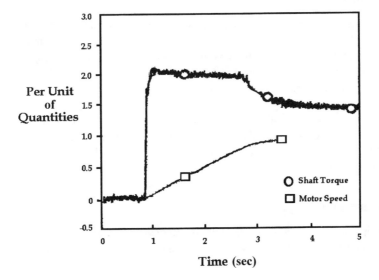

Figure 16–6 Flux-vector inverter response characteristics.

INSTALLATION

Inverters must be installed in a reasonably clean, air-conditioned environment. In critical applications, an extra cooling fan may be required to ensure adequate temperature control. Wiring from an inverter to a motor, including starter, plugsets, and fuse disconnections, must be new or in good working condition.

Motors

Motors to be controlled by an inverter must be of the three-phase, AC type. More than one motor can be controlled by one inverter, but each motor must have its own set of three fuses. A positive disconnection must be installed upstream of the dedicated motor-fuse set.

Short-Circuit Protection

Short-circuit fuse coordination is required in all multimotor inverter systems. In the case of a short circuit in one motor, the individual fuse set will clear without causing the inverter's overcurrent-protection system (fuses, electronic trips, etc.) to operate.

Ground-Fault Protection

Ground-fault fuse coordination is required with multimotor inverter systems for the same reason as short-circuit fuse coordination. Including ground-fault protection for each motor will prevent shutting down the inverter should one or more motors trip.

Transformers

Transformers may be required when a different input or output voltage is required for proper operation. Input transformers are required when the available input voltage is different than that required by the inverter system or motor. Input isolation transformers also can be used to prevent the electrical noise generated in the inverters from feeding into other equipment.

An output transformer can be used to match the motor's load to the inverter. For example, if the inverter is capable of producing a 480-V output and the motor requires only 290 V at maximum speed, an output transformer can be used to step down the inverter's output voltage.

Instrumentation

Proper operation and maintenance of an inverter system requires adequate feedback to monitor and troubleshoot the system. As a minimum, inverter systems should have digital oscillators and a fault-indication system.

Digital Oscillators

Most inverter systems incorporate a digital oscillator. The feedback provided by this device is essential for applications where speed stability is required.

Fault-Indication System

A multimotor inverter system should have sufficient diagnostic capability to detect and display the reason for an unplanned system shutdown. It is desirable for the inverter to display the first fault that occurred.

Other Instrumentation

It is necessary to have voltage, frequency, and current readouts or meters on the front panel of the inverter. Most new inverters have digital displays, which allow the output to be displayed in series or simultaneously.

OPERATING METHODS

Proper operation of an inverter system varies with the type of inverter and the specific application. However, all systems have some common operating parameters, such as startup and braking.

Startup

Motors controlled by inverters must be started across the line, which refers to connecting them across the inverter output while the inverter is operating at running frequency. The inverter must be capable of handling a motor's starting current in-rush without affecting the other motors controlled by the inverter.

Braking

If the application requires the motors to be slowed down faster than they would normally coast to a stop, then dynamic braking is required. This can be achieved with a multimotor inverter system in several ways. When a motor is removed from the inverter output, direct current (DC) from a separate power supply can be connected through one phase of the motor being stopped. Controls must be designed to shut off the DC power when the motor comes to a complete stop. If not, the motor's windings will be damaged.

Another method of braking uses a 60-Hz utility power supply connected to the motor. This 60-Hz power is connected in opposite-phase rotation from the normal direction of the motor. This reverse power provides the braking required to stop the motor. Again, provisions must be made to shut off this power supply as soon as the motor comes to a complete stop.

17

CONTROL VALVES

Control valves can be broken into two major classifications: process and fluid power. Process valves control the flow of gases and liquids through a process system. Fluid-power valves control pneumatic or hydraulic systems.

PROCESS

Process-control valves are available in a variety of sizes, configurations, and materials of construction. Generally, this type of valve is classified by its internal configuration.

Configuration

The device used to control flow through a valve varies with its intended function. The more common types are ball, gate, butterfly, and globe valves.

Ball

Ball valves (see Figure 17–1) are simple shutoff devices that use a ball to stop and start the flow of fluid downstream of the valve. As the valve stem turns to the open position, the ball rotates to a point where part or all of the hole machined through the ball is in line with the valve-body inlet and outlet. This allows fluid to pass through the valve. When the ball rotates so that the hole is perpendicular to the flow path, the flow stops.

Most ball valves are quick-acting and require a 90° turn of the actuator lever to fully open or close the valve. This feature, coupled with the turbulent flow generated when the ball opening is only partially open, limits the use of the ball valve. Use should be limited to strictly an on/off control function (i.e., fully open or fully closed) because of the turbulent flow condition and severe friction loss when in the partially open position. These valves should not be used for throttling or flow control.

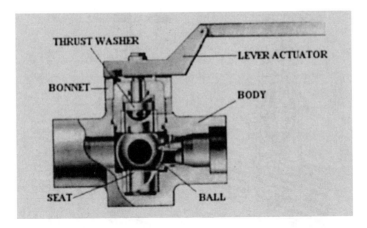

Figure 17–1 Ball valve.

Ball valves used in process applications may incorporate a variety of actuators to provide direct or remote control of the valve. Actuators commonly are either manual or motor operated. Manual values have a handwheel or lever attached directly or through a gearbox to the valve stem. The valve is opened or closed by moving the valve stem through a 90° arc. Motor-controlled valves replace the handwheel with a fractional horsepower motor that can be controlled remotely. The motor-operated valve operates in exactly the same way as the manually operated valve.

Gate

Gate valves are used when straight-line, laminar fluid flow and minimum restrictions are needed. These valves use a wedge-shaped sliding plate in the valve body to stop, throttle, or permit full flow of fluids through the valve. When the valve is wide open, the gate is completely inside the valve bonnet. This leaves the flow passage through the valve fully open, with no flow restrictions, allowing little or no pressure drop through the valve.

Gate valves are not suitable for throttling the flow volume unless specifically authorized for this application by the manufacturer. They generally are not suitable because the flow of fluid through a partially open gate can cause extensive damage to the valve.

Gate valves are classified as either rising stem or non-rising stem. In the non-rising-stem valve, shown in Figure 17–2, the stem is threaded into the gate. As the handwheel on the stem is rotated, the gate travels up or down the stem on the threads, while the stem remains vertically stationary. This type of valve almost always will have a pointer indicator threaded onto the upper end of the stem to indicate the position of the gate.

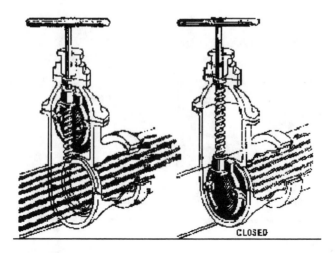

Figure 17–2 Non-rising-stem gate valve (source unknown).

Valves with rising stems (see Figure 17–3) are used when it is important to know by immediate inspection if the valve is open or closed or when the threads exposed to the fluid could become damaged by fluid contamination. In this valve, the stem rises out of the valve bonnet when the valve is opened.

Butterfly

The butterfly valve has a disk-shaped element that rotates about a central shaft or stem. When the valve is closed, the disk face is across the pipe and blocks the flow. Depending on the type of butterfly valve, the seat may consist of a bonded resilient

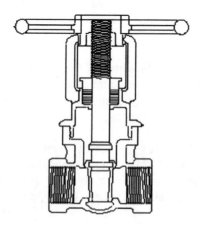

Figure 17–3 Rising stem gate valve.

liner, a mechanically fastened resilient liner, an insert-type reinforced resilient liner, or an integral metal seat with an O-ring inserted around the edge of the disk.

As shown in Figure 17–4, both the fully open and the throttled positions permit almost unrestricted flow. Therefore, this valve does not induce turbulent flow in the partially closed position. While the design does not permit exact flow control, a butterfly valve can be used for throttling flow through the valve. In addition, these valves have the lowest pressure drop of all the conventional types. For such reasons, they commonly are used in process-control applications.

Globe

The globe valve gets its name from the shape of the valve body, although other types of valves also may have globular bodies. Figure 17–5 shows three configurations of this type of valve: straight flow, angle flow and cross flow.

A disk attached to the valve stem controls flow in a globe valve. Turning the valve stem until the disk is seated, illustrated in View A of Figure 17–6, closes the valve. The edge of the disk and the seat are very accurately machined to form a tight seal. It is important for globe valves to be installed with the pressure against the disk face to protect the stem packing from system pressure when the valve is shut.

While this type of valve commonly is used in the fully open or fully closed position, it also may be used for throttling. However, since the seating surface is a relatively large area, it is not suitable for throttling applications where fine adjustments are required.

When the valve is open, as illustrated in View B of Figure 17–6, the fluid flows through the space between the edge of the disk and the seat. Since the fluid flow is equal on all sides of the center of support when the valve is open, no unbalanced pressure is placed

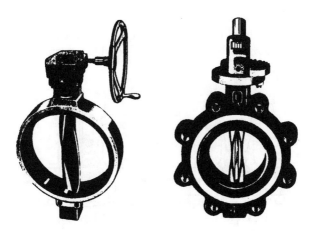

Figure 17–4 Butterfly valves provide almost unrestricted flow (Higgins and Mobley 1995).

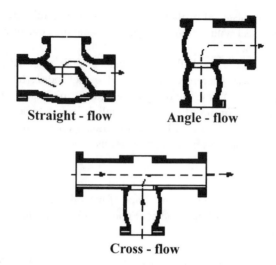

Straight - flow   **Angle - flow**

Cross - flow

Figure 17–5 Three globe valve configurations: straight flow, angle flow, and cross flow.

on the disk to cause uneven wear. The rate at which fluid flows through the valve is regulated by the position of the disk in relation to the valve seat.

The globe valve should never be jammed in the open position. After a valve is fully opened, the handwheel or actuating handle should be closed approximately one-half turn. If this is not done, the valve may seize in the open position making it difficult, if not impossible, to close the valve. Many valves are damaged in the manner. Another reason to partially close a globe valve is because it can be difficult to tell if the valve is open or closed. If jammed in the open position, the stem can be damaged or broken by someone who thinks the valve is closed.

Performance

Process-control valves have few measurable criteria that can be used to determine their performance. Obviously, the valve must provide a positive seal when closed.

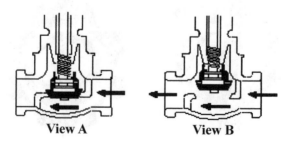

View A   **View B**

Figure 17–6 Globe valve.

In addition, it must provide a relatively laminar flow with minimum pressure drop in the fully open position. When evaluating valves, the following criteria should be considered: capacity rating, flow characteristics, pressure drop, and response characteristics.

Capacity Rating

The primary selection criteria of a control valve is its capacity rating. Each type of valve is available in a variety of sizes to handle most typical process-flow rates. However, proper size selection is critical to the performance characteristics of the valve and the system where it is installed. A valve's capacity must accommodate variations in viscosity, temperature, flow rates, and upstream pressure.

Flow Characteristics

The internal design of process-control valves has a direct impact on the flow characteristics of the gas or liquid flowing through the valve. A fully open butterfly or gate valve provides a relatively straight, obstruction-free flow path. As a result, the product should not be affected. Refer to the previous section on valve configuration for a discussion of the flow characteristics by valve type.

Pressure Drop

The control-valve configuration affects the resistance to flow through the valve. The amount of resistance, or pressure drop, will vary greatly, depending on type, size, and position of the valve's flow-control device (i.e., ball, gate, or disk). Pressure-drop formulas can be obtained for all common valve types from several sources.

Response Characteristics

With the exception of simple, manually controlled shutoff valves, process-control valves generally are used to control the volume and pressure of gases or liquids within a process system. In most applications, valves are controlled from a remote location through the use of pneumatic, hydraulic, or electronic actuators. Actuators are used to position the gate, ball, or disk that starts, stops, directs, or proportions the flow of gas or liquid through the valve. Therefore, the response characteristics of a valve are determined, in part, by the actuator. Three factors critical to proper valve operation are response time, length of travel, and repeatability.

Response Time Response time is the total time required for a valve to open or close to a specific set-point position. These positions are fully open, fully closed, and any position in between. The selection and maintenance of the actuator used to control process-control valves have a major impact on response time.

Length of Travel The valve's flow-control device (i.e., gate, ball, or disk) must travel some distance when going from one set point to another. With a manually operated valve, this is a relatively simple operation. The operator moves the stem lever or handwheel until the desired position is reached. The only reasons why a manually controlled valve will not position properly are mechanical wear or looseness between the lever or handwheel and the disk, ball, or gate.

For remotely controlled valves, however, other variables have a direct impact on valve travel. These variables depend on the type of actuator used. There are three major types of actuators: pneumatic, hydraulic, and electronic.

Pneumatic actuators, including diaphragms, air motors, and cylinders, are suitable for simple on/off valve applications. As long as there is enough air volume and pressure to activate the actuator, the valve can be repositioned over its full length of travel. However, when the air supply required to power the actuator is inadequate or the process-system pressure is too great, the actuator's ability to operate the valve properly is severely reduced.

A pneumatic (i.e., compressed-air driven) actuator is shown in Figure 17–7. This type is not suited for precision flow-control applications, because the compressibility of air prevents it from providing smooth, accurate valve positioning.

Hydraulic (i.e., fluid-driven) actuators, also illustrated in Figure 17–7, can provide a positive means of controlling process valves in most applications. Properly installed and maintained, this type of actuator can provide accurate, repeatable positioning of the control valve over its full range of travel.

Some control valves use high-torque electric motors as their actuator (see Figure 17–8). If the motors are properly sized and their control circuits maintained, this type of actuator can provide reliable, positive control over the full range of travel.

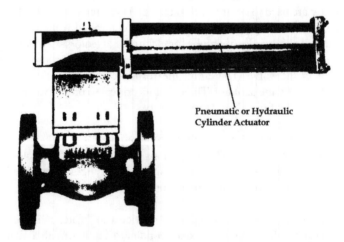

Pneumatic or Hydraulic Cylinder Actuator

Figure 17–7 Pneumatic or hydraulic cylinders are used as actuators (Higgins and Mobley 1995).

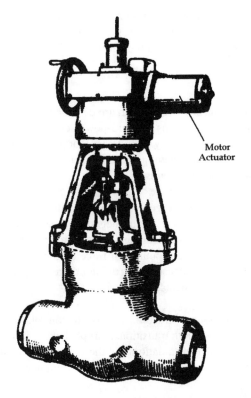

Figure 17–8 High-torque electric motors can be used as actuators (Higgins and Mobley 1995).

Repeatability Repeatability, perhaps, is the most important performance criteria of a process-control valve. This is especially true in applications where precise flow or pressure control is needed for optimum performance of the process system.

New process-control valves generally provide the repeatability required. However, proper maintenance and periodic calibration of the valves and their actuators are required to ensure long-term performance. This is especially true for valves that use mechanical linkages as part of the actuator assembly.

Installation

Process-control valves cannot tolerate solids, especially abrasives, in the gas or liquid stream. In applications where high concentrations of particulates are present, valves tend to experience chronic leakage or seal problems because the particulate matter prevents the ball, disk, or gate from completely closing against the stationary surface.

Simply installing a valve with the same inlet and discharge size as the piping used in the process is not acceptable. In most cases, the valve must be larger than the piping to compensate for flow restrictions within the valve.

Operating Methods

Operating methods for control valves, which are designed to control or direct gas and liquid flow through process systems or fluid-power circuits, range from manual to remote, automatic operation. The key parameters that govern the operation of valves are the speed of the control movement and the impact of speed on the system. This is especially important in process systems.

Hydraulic hammer, the shock wave generated by the rapid change in the flow rate of liquids within a pipe or vessel, has a serious, negative impact on all components of the process system. For example, instantaneously closing a large flow-control valve may generate in excess of 3 million foot-pounds of force on the entire system upstream of the valve. This shock wave can cause catastrophic failure of upstream valves, pumps, welds, and other system components.

Changes in flow rate, pressure, direction, and other controllable variables must be gradual enough to permit a smooth transition. Abrupt changes in valve position should be avoided. Neither the valve installation nor the control mechanism should permit complete shutoff, referred to as *deadheading*, of any circuit in a process system.

Restricted flow forces system components, such as pumps, to operate outside of their performance envelope. This reduces equipment reliability and sets the stage for catastrophic failure or abnormal system performance. In applications where radical changes in flow are required for normal system operation, control valves should be configured to provide an adequate bypass for surplus flow in order to protect the system.

For example, systems that must have close control of flow should use two proportioning valves that act in tandem to maintain a balanced hydraulic or aerodynamic system. The primary, or master, valve should control flow to the downstream process. The second valve, slaved to the master, should divert excess flow to a bypass loop. This master-slave approach ensures that the pumps and other upstream system components are permitted to operate well within their operating envelope.

FLUID POWER

Fluid power control valves are used on pneumatic and hydraulic systems or circuits.

Configuration

The configuration of fluid power control valves varies with their intended application. The more common configurations include one way, two way, three way, and four way.

One Way

One-way valves typically are used for flow and pressure control in fluid-power cir-cuits (see Figure 17–9). Flow-control valves regulate the flow of hydraulic fluid or gases in these systems. Pressure-control valves, in the form of regulators or relief valves, control the amount of pressure transmitted downstream from the valve. In most cases, the types of valves used for flow control are smaller versions of the types of valves used in process control. The major types of process-control valves were dis-cussed previously. These include ball, gate, globe, and butterfly valves.

Pressure-control valves have a third port to vent excess pressure and prevent it from affecting the downstream piping. The bypass, or exhaust, port has an internal flow-control device, such as a diaphragm or piston, that opens at predetermined set points to permit the excess pressure to bypass the valve's primary discharge. In pneumatic circuits, the bypass port vents to the atmosphere. In hydraulic circuits, it must be con-nected to a piping system that returns to the hydraulic reservoir.

Two Way

A two-way valve has two functional flow-control ports. A two-way, sliding spool directional control valve is shown in Figure 17–10. As the spool moves back and forth, it either allows fluid to flow through the valve or prevents it from flowing. In the open position, the fluid enters the inlet port, flows around the shaft of the spool, and through the outlet port. Because the forces in the cylinder are equal when open, the spool cannot move back and forth. In the closed position, one of the spool's pistons simply blocks the inlet port, which prevents flow through the valve.

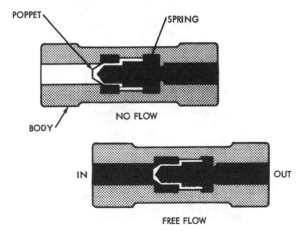

Figure 17–9 One-way, fluid-power valve.

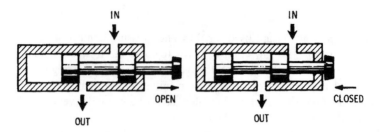

Figure 17–10 Two-way, fluid-power valve (Nelson 1986).

A number of features common to most sliding-spool valves are shown in Figure 17–10. The small ports at either end of the valve housing provide a path for fluid that leaks past the spool to flow to the reservoir. This prevents pressure from building up against the ends of the pistons, which would hinder the movement of the spool. When these valves become worn, they may lose balance because of greater leakage on one side of the spool than on the other. This can cause the spool to stick as it attempts to move back and forth. Therefore, small grooves are machined around the sliding surface of the piston. In hydraulic valves, leaking liquid encircles the piston, keeping the contacting surfaces lubricated and centered.

Three Way

Three-way valves contain a pressure port, cylinder port, and return or exhaust port (see Figure 17–11). The three-way directional control valve is designed to operate an

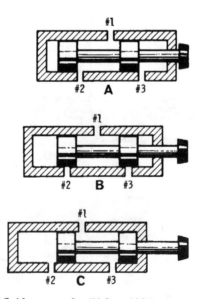

Figure 17–11 Three-way, fluid-power valve (Nelson 1986).

actuating unit in one direction. It is returned to its original position either by a spring or the load on the actuating unit.

Four Way

Most actuating devices require system pressure in order to operate in two directions. The four-way directional control valve, which contains four ports, is used to control the operation of such devices (see Figure 17–12). The four-way valve also is used in some systems to control the operation of other valves. It is one of the most widely used directional-control valves in fluid-power systems.

The typical four-way directional control valve has four ports: a pressure port, a return port, and two cylinder or work (output) ports. The pressure port is connected to the main system-pressure line and the return port is connected to the reservoir return line. The two outputs are connected to the actuating unit.

Performance

The criteria that determines performance of fluid-power valves are similar to those for process-control valves as discussed previously. As with process-control valves, fluid-power valves must be selected based on their intended application and function.

Installation

When installing fluid power control valves, piping connections are made either directly to the valve body or to a manifold attached to the valve's base. Care should be taken to ensure that the piping is connected to the proper valve port. The schematic diagram affixed to the valve body will indicate the proper piping arrangement, as well

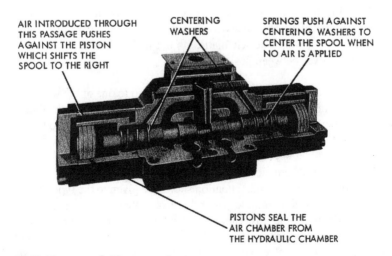

Figure 17–12 Four-way, fluid-power valves.

as the designed operation of the valve. In addition, the ports on most fluid-power valves generally are clearly marked to indicate their intended function.

In hydraulic circuits, the return or common ports should be connected to a return line that directly connects the valve to the reservoir tank. This return line should not need a pressure-control device but should have a check valve to prevent reverse flow of the hydraulic fluid.

Pneumatic circuits may be vented directly to atmosphere. A return line can be used to reduce noise or any adverse effect that locally vented compressed air might have on the area.

Operating Methods

The function and proper operation of a fluid-power valve are relatively simple. Most of these valves have a schematic diagram affixed to the body that clearly explains how to operate the valve.

Backup Valves

Figure 17–13 is a schematic of a two-position, cam-operated valve. The primary actuator, or cam, is positioned on the left of the schematic and any secondary actuators are on the right. In this example, the secondary actuator consists of a spring-return and a spring-compensated limit switch. The schematic indicates that, when the valve is in the neutral position (right box), flow is directed from the inlet (P) to work port A. When the cam is depressed, the flow momentarily shifts to work port B. The secondary actuator, or spring, automatically returns the valve to its neutral position when the cam returns to its extended position. In these schematics, T indicates the return connection to the reservoir.

Figure 17–14 illustrates a typical schematic of a two-position and three-position directional control valve. The boxes contain flow direction arrows that indicate the flow path in each position. The schematics do not include the actuators used to activate or shift the valves between positions.

In a two-position valve, the flow path is always directed to one of the work ports (A or B). In a three-position valve, a third or neutral position is added. In this figure, a Type 2 center position is used. In the neutral position, all ports are blocked and no flow through the valve is possible.

Figure 17–15 is the schematic for the center or neutral position of three-position directional control valves. Special attention should be given to the type of center position that is used in a hydraulic control valve. When Type 2, 3, and 6 (see Figure 17–15) are used, the upstream side of the valve must have a relief or bypass valve installed. Since the pressure port is blocked, the valve cannot relieve pressure on the upstream side of the valve. The Type 4 center position, called a *motor spool*, per-

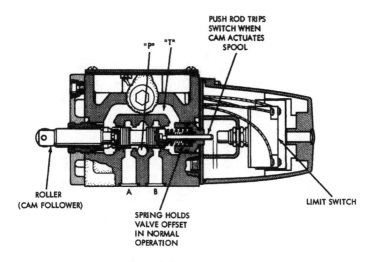

Figure 17–13 Schematic for a cam-operated, two-position valve.

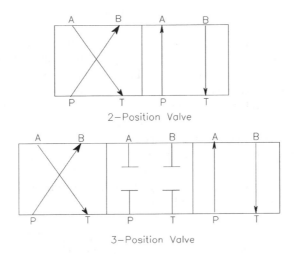

Figure 17–14 Schematic of two-position and three-position valves.

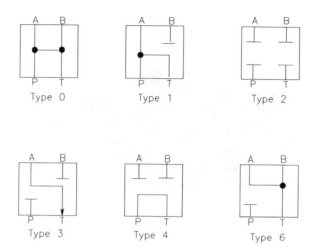

Figure 17–15 Schematic for center or neutral configurations of three-position valves.

mits the full pressure and volume on the upstream side of the valve to flow directly to the return line and storage reservoir. This is the recommended center position for most hydraulic circuits.

The schematic affixed to the valve includes the primary and secondary actuators used to control the valve. Figure 17–16 provides the schematics for three actuator-controlled valves:

1. Double-solenoid, spring-centered, three-position valve;
2. Solenoid-operated, spring-return, two-position valve;
3. Double-solenoid, detented, two-position valve.

The top schematic represents a double-solenoid, spring-centered, three-position valve. When neither of the two solenoids is energized, the double springs ensure that the valve is in its center or neutral position. In this example, a Type 0 (see Figure 17–15) configuration is used. This neutral-position configuration equalizes the pressure through the valve. Since the pressure port is open to both work ports and the return line, pressure is equalized throughout the system. When the left or primary solenoid is energized, the valve shifts to the left-hand position and directs pressure to work port B. In this position, fluid in the A side of the circuit returns to the reservoir. As soon as the solenoid is de-energized, the valve shifts back to the neutral or center position. When the secondary (i.e., right) solenoid is energized, the valve redirects flow to port A and port B returns fluid to the reservoir.

The middle schematic represents a solenoid-operated, spring-return, two-position valve. Unless the solenoid is energized, the pressure port (P) is connected to work port A. While the solenoid is energized, flow is redirected to work port B. The spring

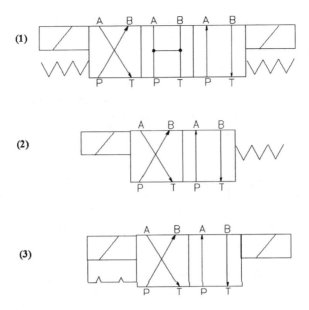

Figure 17–16 Actuator-controlled valve schematics.

return ensures that the valve is in its neutral (i.e., right) position when the solenoid is de-energized.

The bottom schematic represents a double-solenoid, detented, two-position valve. The solenoids are used to shift the valve between its two positions. A secondary device, called a *detent*, is used to hold the valve in its last position until the alternate solenoid is energized. Detent configuration varies with the valve type and manufacturer. However, all configurations prevent the valve's control device from moving until a strong force, such as that provided by the solenoid, overcomes its locking force.

Actuators

As with process-control valves, actuators used to control fluid-power valves have a fundamental influence on performance. The actuators must provide positive, real-time response to control inputs. The primary types of actuators used to control fluid-power valves are mechanical, pilot, and solenoid.

Mechanical The use of manually controlled mechanical valves is limited in both pneumatic and hydraulic circuits. Generally, this type of actuator is used only on isolation valves that are activated when the circuit or fluid-power system is shut down for repair or when direct operator input is required to operate one of the system components.

Manual control devices (e.g., levers, cams, or palm buttons) can be used as the primary actuator on most fluid power control valves. Normally, these actuators are used in conjunction with a secondary actuator, such as a spring return or detent, to ensure proper operation of the control valve and its circuit.

Spring returns are used in applications where the valve is designed to stay open or shut only when the operator holds the manual actuator in a particular position. When the operator releases the manual control, the spring returns the valve to the neutral position.

Valves with a detented secondary actuator are designed to remain in the last position selected by the operator until manually moved to another position. A detent actuator is simply a notched device that locks the valve in one of several preselected positions. When the operator applies force to the primary actuator, the valve shifts out of the detent and moves freely until the next detent is reached.

Pilot Although a variety of pilot actuators is used to control fluid-power valves, they all work on the same basic principle. A secondary source of fluid or gas pressure is applied to one side of a sealing device, such as a piston or diaphragm. As long as this secondary pressure remains within preselected limits, the sealing device prevents the control valve's flow-control mechanism (i.e., spool or poppet) from moving. However, if the pressure falls outside the preselected window, the actuator shifts and forces the valve's primary mechanism to move to another position.

This type of actuator can be used to sequence the operation of several control valves or operations performed by the fluid-power circuit. For example, a pilot-operated valve is used to sequence the retraction of an airplane's landing gear. The doors that conceal the landing gear when retracted cannot close until the gear is fully retracted. A pilot-operated valve senses the hydraulic pressure in the gear-retraction circuit. When the hydraulic pressure reaches a preselected point that indicates the gear is fully retracted, the pilot-actuated valve triggers the closure circuit for the wheel-well doors.

Solenoid Solenoid valves are widely used as actuators for fluid-power systems. This type of actuator consists of a coil that generates an electric field when energized. The magnetic forces generated by this field force a plunger attached to the main valve's control mechanism to move within the coil. This movement changes the position of the main valve.

In some applications, the mechanical force generated by the solenoid coil is not sufficient to move the main valve's control mechanism. When this occurs, the solenoid actuator is used in conjunction with a pilot actuator. The solenoid coil opens the pilot port, which uses system pressure to shift the main valve.

Solenoid actuators always are used with a secondary actuator to provide positive control of the main valve. Because of heat buildup, solenoid actuators must be limited to short-duration activation. A brief burst of electrical energy is transmitted to the solenoid's coil and the actuation triggers a movement of the main valve's control mechanism. As soon as the main valve's position is changed, the energy to the solenoid coil is shut off.

This operating characteristic of solenoid actuators is important. For example, a normally closed valve that uses a solenoid actuation can be open only when the solenoid is energized. As soon as the electrical energy is removed from the solenoid's coil, the valve returns to the closed position. The reverse is true of a normally open valve. The main valve remains open, except when the solenoid is energized.

The combination of primary and secondary actuators varies with the specific application. Secondary actuators can be another solenoid or any of the other actuator types that have been previously discussed.

18

SEALS AND PACKING

All machines that handle liquids or gases, such as pumps and compressors, must be sealed around their shafts to prevent fluid from leaking. To accomplish this, the machine design must include seals located at various points to prevent leakage between the shaft and housing. This chapter discusses sealing requirements and common seals.

CONFIGURATION

The two primary types of sealing devices used to seal around rotating shafts are packed-stuffing boxes and simple mechanical seals.

Packed-Stuffing Boxes

A soft, pliable packing material placed in a box and compressed into rings encircling the drive shaft commonly is used to prevent leakage. Packing rings between the pump housing and the drive shaft, compressed by tightening the gland-stuffing follower, forms an effective seal. Figure 18–1 shows a typical packed-stuffing box seal.

Simple Mechanical Seal

A mechanical seal is used on centrifugal pumps and other types of fluid-handling equipment where shaft sealing is critical and no leakage can be tolerated. Toxic chemicals and other hazardous materials are primary examples of applications where mechanical seals are used. These seals also are referred to as *friction drives*, or *single-coil spring seals*, and *positive drives*.

Figure 18–2 shows the components of a simple mechanical seal, which is made up of a coil spring, O-ring shaft packing, and a seal ring. The seal ring fits over the shaft and

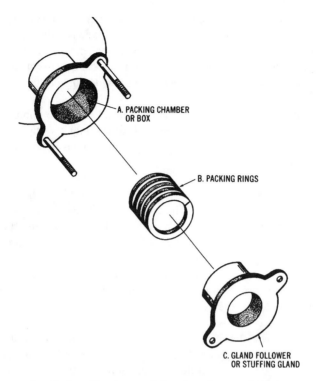

Figure 18–1 Typical packed-stuffing box seal (Bearings Inc. catalogue).

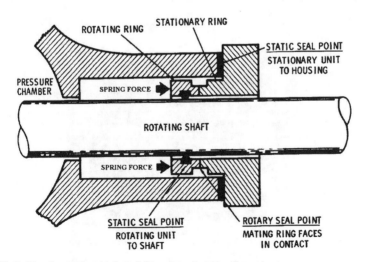

Figure 18–2 Simple mechanical seal (Bearings Inc. catalogue).

rotates with it. The spring must be made from a material compatible with the fluid being pumped so that it will withstand corrosion. Likewise, the same care must be taken with material selection for the O ring and seal materials. A carbon graphite insertion ring provides a good bearing surface for the seal ring to rotate against. It also is resistant to attack by corrosive chemicals over a wide range of temperatures.

Figure 18–3 shows a simple seal that has been installed in the pump's stuffing box. Note how the coil spring sits against the back of the pump's impeller, pushing the packing O ring against the seal ring. By doing so, it remains in constant contact with the stationary insert ring.

As the shaft rotates, the packing rotates with it, due to friction. There also is friction between the spring, the impeller, and the compressed O ring. Thus, the whole assembly rotates together when the pump's shaft rotates. The stationary insert ring is located within the gland bore. The gland itself is bolted to the face of the stuffing box. This part is held stationary by the friction between the O ring insert mounting and the inside diameter (I.D.) of the gland bore as the shaft rotates within the bore of the insert. In more complex mechanical seals, the shaft packing element can be secured to the rotating shaft by Allen screws.

Having discussed how a simple mechanical seal is assembled in the stuffing box, we must now consider how the pumped fluid is prevented from leaking out to the atmosphere. In Figure 18–3, the path of the fluid along the drive shaft is blocked by the O ring shaft packing at Point A. Any fluid attempting to pass through the seal ring is stopped by the O ring shaft packing at Point B. Any further attempt by the fluid to pass through the seal ring to the atmospheric side of the pump is prevented by the gland gasket at Point C and the O-ring insert at Point D. The only other place where fluid can escape is the joint surface around Point E, which is between the rotating car-

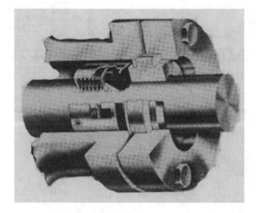

Figure 18–3 Pump stuffing box containing a simple mechanical seal (Bearings Inc. catalogue).

bon ring and the stationary insert. Note that the surface areas of both rings must be machined-lapped perfectly flat, measured in angstroms with tolerances of one-millionth of an inch.

PERFORMANCE

Performance of a packed-stuffing box seal depends primarily on the presence of a small quantity of fluid through the box. This flow is needed to provide both lubrication and cooling of the packing.

A mechanical seal's performance depends on the operating condition of the equipment on which it is installed. Its efficiency depends on the condition of the sealing-area surfaces, which are friction-bearing surfaces that remain in contact with each other for the effective working life of the seal.

This type of seal is more reliable than compressed packing seals. Because the spring in a mechanical seal exerts constant pressure on the seal ring, it automatically adjusts for wear at the faces. Therefore, the need for manual adjustment is eliminated. Additionally, because the bearing surface is between the rotating and stationary components of the seal, the shaft or shaft sleeve does not become worn. Although the seal eventually will wear out and need replacing, the shaft will not experience wear.

INSTALLATION

This section describes the installation procedures for packed-stuffing boxes and mechanical seals.

Packed-Stuffing Box

The following sections provide detailed instructions on how to repack centrifugal pump packed-stuffing boxes or glands. The methodology described here applies to other gland-sealed units, such as valves and reciprocating machinery.

Tool List

The following list specifies the tools needed to repack a centrifugal pump gland:

- Approved packing for specific equipment,
- Mandrel sized to shaft diameter,
- Packing ring extractor tool,
- Packing board,
- Sharp knife,
- Approved cleaning solvent,
- Lint-free cleaning rags.

Precautions

The following precautions should be taken when repacking a packed-stuffing box:

- Coordinate with operations control.
- Observe site and area safety precautions at all times.
- Ensure that equipment has been electrically isolated and suitably locked out and tagged.

Ensure machine is isolated and depressurized with suction and discharge valves chained and locked shut.

Installation

The following steps are followed when installing a gland:

1. Loosen and remove nuts from the gland bolts.
2. Examine threads on bolts and nuts for stretching or damage; replace if defective.
3. Remove the gland follower from the stuffing box and slide it along the shaft to provide access to the packing area.
4. Use a packing extraction tool to carefully remove packing from the gland.
5. Keep the packing rings in the order they are removed from the gland box. This is important in evaluating wear characteristics. Look for rub marks and any other unusual markings that would identify operational problems.
6. Carefully remove the lantern ring. This is a grooved, bobbinlike spool situated exactly on the centerline of the seal water inlet connection to the gland (Figure 18–4). NOTE: It is most important to place the lantern ring under the seal water inlet connection to ensure that the water is properly distributed within the gland to perform its cooling and lubricating functions.
7. Examine the lantern ring for scoring and possible signs of crushing. Make sure the lantern ring's outside diameter (O.D.) provides a sliding fit with the gland box's internal dimension. Check that the lantern ring's I.D. is a free fit along the pump's shaft sleeve. If the lantern ring does not meet this simple criterion, replace it with a new one.

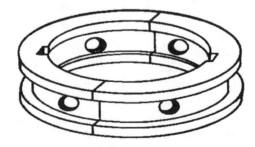

Figure 18–4 Lantern ring or seal cage (95/96 Product Guide).

8. Continue to remove the rest of the packing rings as previously described. Keep each ring and note the sequence that it was removed.
9. Do not discard packing rings until they have been thoroughly examined for potential problems.
10. Turn on the gland seal cooling water slightly to ensure there is no blockage in the line. Shut the valve when good flow conditions are established.
11. Repeat Steps 1 through 10 with the other gland box.
12. Carefully clean out the gland-stuffing boxes with a solvent-soaked rag to ensure that no debris is left behind.
13. Examine the shaft sleeve in both gland areas for excessive wear caused by poorly lubricated or overtightened packing. NOTE: If the shaft sleeve is ridged or badly scratched in any way, the split housing of the pump may have to be taken apart and the impeller removed for the sleeve to be replaced. This is caused by badly installed and maintained packing.
14. Check total indicated runout (TIR) of the pump shaft by placing a magnetic base-mounted dial indicator on the pump housing and a dial stem on the shaft. Turn the dial to 0 and rotate the pump shaft one full turn. Record the reading (see Figure 18–5). NOTE: If the TIR is greater than ±0. 002 in., the pump shaft should be straightened.
15. Determine the correct packing size before installing using the following method: Referring to Figure 18–6, measure the I.D. of the stuffing box, which is the O.D. at the packing (B), and the diameter of the shaft (A). With this data, the packing cross-section size is calculated by

$$\text{Packing Cross-Section} = \frac{B - A}{2}$$

The packing length (PL) is determined by calculating the circumference of the packing within the stuffing box. The centerline diameter is calculated

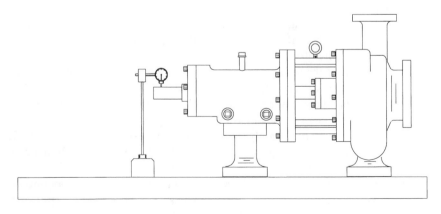

Figure 18–5 Dial indicator check for runout (Bearings Inc. catalogue).

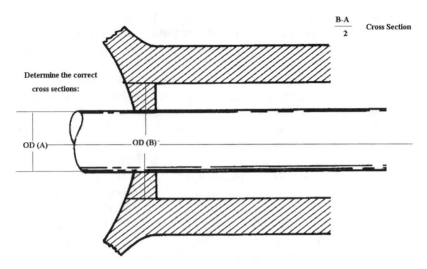

Figure 18–6 Selecting correct packing size (Bearings Inc. catalogue).

by adding the diameter of the shaft to the packing cross-section that was calculated in the preceding formula. For example, a stuffing box with a 4-in. I.D. and a shaft with a 2-in. diameter will require a packing cross-section of 1 in. The centerline of the packing then would be 3 in. Therefore, the approximate length of each piece of packing would be

$$PL = \text{Centerline diameter} \times 3.1416 = 3.0 \times 3.1416 = 9.43 \text{ in.}$$

The packing should be cut approximately $\frac{1}{4}$ in. longer than the calculated length so the end can be bevel cut.

16. Controlled leakage rates easily can be achieved with the correct size of packing.
17. Cut the packing rings to size on a wooden mandrel that is the same diameter as the pump shaft. Rings can be cut either square (butt cut) or diagonally at approximately 30°. NOTE: Leave at least a $\frac{1}{16}$-in. gap between the butts regardless of the type of cut used. This permits the packing rings to move under compression or temperature without binding on the shaft surface.
18. Ensure that the gland area is perfectly clean and not scratched in any way before installing the packing rings.
19. Lubricate each ring lightly before installing in the stuffing box. NOTE: When putting packing rings around the shaft, use an S twist. Do not bend open. See Figure 18–7.
20. Use a split bushing to install each ring, ensuring that the ring bottoms out inside the stuffing box. An offset tamping stick may be used if a split bushing is not available. Do not use a screwdriver.

"S" Twist

Wrong

Figure 18–7 Proper and improper installation of packing (Bearings Inc. catalogue).

21. Stagger the butt joints, placing the first ring butt at 12 o'clock; the second at 6 o'clock; the third at 3 o'clock; the fourth at 9 o'clock; and so forth, until the packing box is filled (Figure 18–8). NOTE: When the last ring has been installed, there should be enough room to insert the gland follower $\frac{1}{8}$ to $\frac{1}{16}$ in. into the stuffing box (Figure 18–9).
22. Install the lantern ring in its correct location within the gland. Do not force the lantern ring into position (Figure 18–10).

Figure 18–8 Stagger butt joints (Bearings Inc. catalogue).

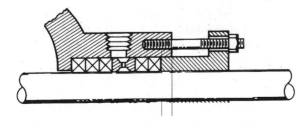

Figure 18–9 Proper gland follower clearance (Bearings Inc. catalogue).

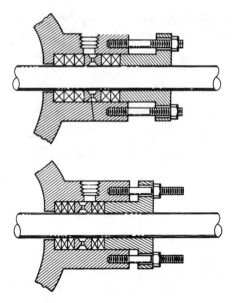

Figure 18–10 Proper lantern ring installation (Bearings Inc. catalogue).

23. Tighten up the gland bolts with a wrench to seat and form the packing to the stuffing box and shaft.
24. Loosen the gland nuts one complete turn and rotate the shaft by hand to get running clearance.
25. Retighten the nuts finger tight only. Again, rotate the shaft by hand to make sure the packing is not too tight.
26. Have the operator start the pump and allow the stuffing box to leak freely. Tighten the gland bolts one flat at a time until the desired leakage is obtained and the pump runs cool.
27. Clean up the work area and account for all tools before returning them to the tool crib.
28. Inform operations of project status and complete all paperwork.
29. After the pump is in operation, periodically inspect the gland to check its performance. If it tends to leak more than the allowable amount, tighten by turning the nuts one flat at a time. Give the packing enough time to adjust before tightening it more. If the gland is tightened too much at one time, the packing can be compressed excessively, which can cause unnecessary friction and subsequent packing burnout.

Mechanical Seals

Much more precision and attention to detail must be given to the installation of mechanical seals than to packing. Nevertheless, it is not unusual for mechanical seals to remain in service for many thousands of operational hours if they have been properly installed and maintained. Inspection of the equipment before seal installation can prevent potential seal failure and reduce overall maintenance expenses.

Equipment Checkpoints

The preinstallation equipment inspection should include the following: stuffing-box space, lateral or axial shaft movement (end play), radial shaft movement (whip or deflection), shaft runout (bent shaft), stuffing box face squareness, stuffing box bore concentricity, driver alignment, and pipe strain.

Stuffing-Box Space To properly receive the seal, the radial space and depth of the stuffing box must be the same as the dimensions shown on the seal's assembly drawing.

End Play A mechanical seal cannot work properly with a large amount of end play or lateral movement. If the hydraulic condition changes (as frequently happens), the shaft could "float," resulting in sealing problems. Minimum end play is a desirable condition for the following reasons:

- Excessive end play can cause pitting, fretting, or wear at the point of contact between the shaft packing in the mechanical seal and the shaft or sleeve O.D. As the mechanical seal driving element is locked to the shaft or sleeve, any excessive end play will result in either overloading or underloading the springs, causing excessive wear or leaks.
- Excessive end play as a result of defective thrust bearings can reduce seal performance by disturbing both the established wear pattern and the lubricating film.
- A floating shaft can cause chattering, which results in seal-face chipping, especially the carbon element. Ideal mechanical seal performance requires a uniform wear pattern and a liquid film between the mating contact faces.

Measure end play by installing a dial indicator with the stem against the shoulder of the shaft. Use a soft hammer or mallet to lightly tap the shaft on one end and then on the other. Total indicated end play should be between 0.001 and 0.004 in.

Whip or Deflection Install the dial indicator as close to the radial bearing as possible. Lift the shaft or exert light pressure at the impeller end. If more than 0.002 to 0.003 in. of radial movement occurs, investigate bearings for radial bearing fit (especially the bore). An oversized radial bearing bore caused by wear, improper machining, or corrosion will cause excessive radial shaft movement, resulting in shaft whip and deflection. Minimum radial shaft movement is important for the following reasons:

- Excessive radial movement can cause wear, fretting, or pitting of the shaft packing or secondary sealing element at the point of contact between the shaft packing and the shaft or sleeve O.D.
- Extreme wear at the mating contact faces will occur when excessive shaft whip or deflection is present due to defective radial bearings or bearing fits.

The contact area of the mating faces will be increased, resulting in increased wear and the elimination or reduction of the lubricating film between the faces, further shortening seal life.

Shaft Runout   *Shaft runout* refers to a bent shaft, which can lead to vibration and poor sealing. Bearing life is greatly reduced, and the operating conditions of both radial and thrust bearings can be affected.

Measure the shaft runout at two or more points on the O.D. of the impeller end of the shaft by clamping the dial indicator to the pump housing. Also measure the shaft runout at the coupling end of the shaft. If the measurement exceeds 0.002 in., straighten or replace the shaft.

Face Squareness   With the stuffing-box cover bolted down, clamp the dial indicator to the shaft with the stem against the face of the box. The total indicator runout should not exceed 0.003 in. When the face of the stuffing box is "out of square," not perpendicular to the shaft axis, the result can be serious malfunction of a mechanical seal for the following reasons:

- The stationary gland plate that holds the stationary insert or seat in position is bolted to the face of the stuffing box. Misalignment will cause the gland to cock, causing the stationary element to cock. This results in seal wobble or operation in an elliptical pattern. This condition is a major factor in fretting, pitting, and wearing of the mechanical seal shaft packing at the point of contact with the shaft or sleeve.
- A seal that wobbles on the shaft can cause wear on the drive pins. Erratic wear on the face contact results in poor seal performance.

Bore Concentricity   With the dial indicator set up as described previously, place the indicator stem well into the bore of the stuffing box, which should be concentric to the shaft axis to within a 0.005 in. total indicator reading. Eccentricity alters the hydraulic loading of the seal faces, reducing seal life and performance. If the shaft is eccentric to the box bore, check the slop, or looseness, in the pump-bracket fits. Rust, atmospheric corrosion, or corrosion from leaking gaskets can cause damage to these fits, making it impossible to ensure that a stuffing box is concentric with the shaft. A possible remedy for this condition is welding the corroded area and remachining it to proper dimensions.

Driver Alignment and Pipe Strain   Driver alignment is extremely important, and periodic checks should be performed. Pipe strain also can damage pumps, bearings, and seals. In most plants, it is customary to blind the suction and discharge flanges of inactive pumps. These blinds should be removed before pump-driver alignment is performed or the job is incomplete.

After the blinds have been removed and as the flanges on the suction and discharge are being connected to the piping, check the dial-indicator reading on the O.D. of the coupling half. Observe movement of the indicator dial as the flanges are being secured. Deviation indicates pipe strain. If severe strain exists, corrective measures should be taken or damage to the pump and unsatisfactory seal service can result.

Seal Checkpoints

The following are important seal checkpoints:

- Ensure that all parts are kept clean, especially the running faces of the seal ring and insert.
- Check the seal rotary unit and make sure the drive pins and spring pins are free in the pin holes or slots.
- Check the set screws in the rotary unit collar to see that they are free in the threads. Set screws should be replaced after each use.
- Check the thickness of all gaskets against the dimensions shown on the assembly drawing. Improper gasket thickness will affect the seal setting and the spring load imposed on the seal.
- Check the fit of the gland ring to the equipment. Make sure there are no obstructions or binding with the studs or bolts. Be sure the gland-ring pilot, if any, enters the bore with a reasonable guiding fit for proper seal alignment.
- Make sure all rotary-unit parts of the seal fit over the shaft freely.
- Check both running faces of the seal (seal ring and insert) and be sure there are no nicks or scratches. Imperfections of any kind on either of these faces will cause leaks.

Installation of the Seal

The following steps should be taken when installing a seal:

- Instruction booklets and a copy of the assembly drawing are generally shipped with each seal. Read the instructions and study the drawing before starting installation.
- Remove all burrs and sharp edges from the shaft or shaft sleeve, including sharp edges of keyways and threads. Worn shafts or sleeves should be replaced.
- Check the stuffing-box bore and face to ensure they are clean and free of burrs.
- Lightly oil the shaft or sleeve before assembling the seal to allow the parts to move freely over it. This is especially desirable when assembling the seal collar because its bore usually has only a few thousandths of an inch clearance. Take care to avoid cocking the collar.
- Install the rotary-unit parts on the shaft or sleeve in the proper order.
- Be careful when passing the insert and gland ring over the shaft. Do not bring the insert against the shaft because it might chip away small pieces from the edge of the running face.
- Wipe the seal faces clean and apply a clean oil film before completing the equipment assembly. A clean finger, which does not leave lint, is best for the final wipe of the seal face.

- Complete the equipment assembly, taking care when compressing the seal into the stuffing box.
- Seat the gland ring and gasket to the face of the stuffing box by tightening the nuts and bolts evenly and firmly. Be sure the gland ring is not cocked. Tighten the nuts and bolts only enough to form a seal at the gland ring gasket, usually finger tight and one half to three quarters of a turn with a wrench. Excessively tightening the gland ring nuts and bolts will cause distortion that will be transmitted to the running face, resulting in leaks.
- If the seal's assembly drawing is not available, the proper setting dimension for inside seals can be determined as follows:
  - Establish a reference mark on the shaft or sleeve flush with the face of the stuffing box.
  - Determine how far the face of the insert will extend into the stuffing-box bore. Take this dimension from the face of the gasket.
  - Determine the compressed length of the rotary unit by compressing it to the proper spring gap.
  - This dimension, added to the distance the insert extends into the stuffing box, gives the seal-setting dimension from the reference mark on the shaft or sleeve to the back of the seal collar.
  - Outside seals are set with the spring gap equal to the dimension stamped on the seal collar.
- Cartridge seals are set at the factory and installed as complete assemblies. These assemblies contain spacers that must be removed after being bolted into position and the sleeve collar is in place.

Installation of Environmental Controls

Mechanical seals often are chosen and designed to operate with environmental controls. If this is the case, check the seal's assembly drawing or the equipment's drawing to ensure that all environmental-control piping is properly installed.

Seal Startup Procedures

Before equipment startup, all heating and cooling lines should be operating. These lines also should remain in operation for a short period after equipment shutdown. On double-seal installations, be sure the liquid lines are connected, the pressure-control valves are properly adjusted, and the sealing-liquid system is operating before starting the equipment.

Before startup, all systems should be properly vented. This is especially important on vertical installations where the stuffing box is the uppermost portion of the pressure-containing part of the equipment. The stuffing-box area must be properly vented to avoid a vapor lock in the seal area that would cause it to run dry.

When starting equipment with mechanical seals, make sure the seal faces are immersed in liquid from the beginning so they will not be damaged from dry operation. The following recommendations for seal startup apply to most types of seal installations and will improve their life if followed:

- Caution the electrician not to run the equipment dry while checking motor rotation. A slight turnover will not hurt the seal, but operating at full speed for several minutes under dry conditions will destroy or severely damage the rubbing faces.
- The stuffing box always should be vented before startup, especially with centrifugal pumps. Even if the pump has a flooded suction, it is still possible that air may be trapped in the top of the stuffing box after the pump's initial liquid purge.
- Where cooling or bypass recirculation taps are incorporated in the seal gland, piping must be connected to or from these taps before startup. These specific environmental-control features must be used to protect the organic materials in the seal and to ensure proper performance.
- Cooling lines should be left open at all times. This is especially true when hot product passes through off-line standby equipment, commonly done so that additional product volume or equipment change can be achieved easily, often by simply pushing a button.
- At the end of each day when hot operational equipment is shut down, it is best to leave the cooling water on long enough for the seal area to cool below the temperature limits of the organic materials in the seal.
- Before startup, face-lubricated seals must be connected from the source of lubrication to the tap openings in the seal gland. For double seals, it is necessary for the lubrication feed lines to be connected to the proper ports before startup for both circulatory and dead-end systems. This is very important because all types of double seals depend on the controlled pressure and flow of the sealing fluid to function properly. Even before the shaft is rotated, the sealing liquid pressure must exceed the product pressure opposing the seal. Be sure a vapor trap does not prevent the lubricant from promptly reaching the seal face.
- Thorough warm-up procedures include a check of all steam piping arrangements to be sure that all are connected and functioning. Products that solidify when cool must be fully melted before startup. It is advisable to leave all heat sources on while the system is shut down to ensure that the product remains in the liquid state. This facilitates quick startups and equipment switchovers that may be required during a production cycle.
- Thorough chilling procedures are necessary for some applications; for example, applications involving liquefied petroleum gas (LPG). LPG always must be kept in a liquid state in the seal area, and startup usually is the most critical time. Even during operation, the recirculation line piped to the stuffing box might need to be run through a cooler to overcome frictional heat generated at the seal faces. LPG requires a stuffing-box pressure greater than the vapor pressure of the product at pumping temperature. A 25 to 50 psi differential is generally desired.

OPERATING METHODS

This section discusses operating methods for packed-stuffing boxes and simple mechanical seals.

Packed-Stuffing Boxes

Packed-stuffing boxes commonly are used on slow- to moderate-speed machinery where a slight amount of leakage is permissible. If the packing is allowed to operate against the shaft without adequate lubrication and cooling, frictional heat eventually will build up to the point of total packing destruction and damage to the drive shaft. Therefore, all packed boxes must have a means of lubrication and cooling.

Lubrication and cooling can be accomplished by allowing a small amount of leakage of fluid from the machine or by providing an external source of fluid. When leakage from the machine is used, leaking fluid is captured in collection basins built into the machine housing or baseplate. Note that periodic maintenance to recompress the packing must be carried out when leakage becomes excessive.

Packed boxes must be protected against ingress of dirt and air, which can result in loss of resilience and lubricity. When this occurs, packing will act like a grinding stone, effectively destroying the shaft's sacrificial sleeve and causing the gland to leak excessively. When the sacrificial sleeve on the drive shaft becomes ridged and worn, it should be replaced as soon as possible. In effect, this is a continuing maintenance program that readily can be measured in terms of dollars and time.

Uneven pressure can be exerted on the drive shaft due to irregularities in the packing rings, resulting in irregular contact with the shaft. This causes uneven distribution of lubrication at certain locations, producing acute wear and packed-box leakage. The only effective solution to this problem is to replace the shaft sleeve or drive shaft at the earliest opportunity.

Simple Mechanical Seal

As with compressed packing glands, lubrication must be provided in mechanical seals. The sealing-area surfaces should be lubricated and cooled with pumped fluid (if it is clean enough) or another source of clean fluid. However, much less lubrication is required with this type of seal because the frictional surface area is smaller than that of a compressed-packing gland and the contact pressure is equally distributed throughout the interface. As a result, a smaller amount of lubrication passes between the seal faces to exit as leakage.

Most packing glands have a measurable flow of lubrication fluid between the packing rings and the shaft. With mechanical seals, the faces ride on a microscopic film of fluid that migrates between them and results in leakage. However, leakage is so slight that, if the temperature of the fluid is above its saturation point at atmospheric pressure, it flashes off to vapor before it can be visually detected.

Friction Drive or Single-Coil Spring Seal

The seal shown back in Figure 18–2 is a typical friction drive, or single-coil spring seal unit. This design is limited in use to nonlubricating fluids (e.g., water) because it relies on friction to turn the rotary unit. For use with liquids that have natural lubricating properties, the seal must be mechanically locked to the drive shaft.

Two drawbacks must be considered for this type of seal. Both are related to the use of a coil spring that fits over the drive shaft. Nevertheless, the simple and reliable coil spring seal has proven itself in the pumping industry and often is specified despite its drawbacks. In regulated industries, this type of seal design far exceeds the capabilities of a compressed packing ring seal.

One drawback of the spring seal is the need for relatively low shaft speeds. The components have a tendency to distort at high surface speeds. This makes the spring push harder on one side of the seal than the other, resulting in an uneven liquid film between the faces, which causes excessive leakage and wear at the seal.

The other drawback is simply one of economics. Because pumps come in a variety of shaft sizes and speeds, the use of this type of seal requires several sizes of spare springs be kept in inventory.

Positive Drive

There are two methods of converting a simple seal to positive drive. Both methods, which use collars secured to the drive shaft by set screws, are shown in Figure 18–11. In the Figure on the left, the end tabs of the spring are bent at 90° to the natural curve of the spring. These end tabs fit into notches in both the collar and the seal ring. This design transmits rotational drive from the collar to the seal ring by the spring. In the right drawing of Figure 18–11, two horizontally mounted pins extend over the spring from the collar to the seal ring.

Figure 18–11 Conversion of a simple seal to positive drive (Roberts 1978).

Part III

EQUIPMENT TROUBLESHOOTING GUIDE

Most machine trains are prone to certain kinds of abnormal behavior and generally exhibit a finite number of recurring failure modes. In most cases, failures result from improper maintenance and operating practices that do not abide by design limits and restrictions.

This part provides an overview of the more common failure modes for machinery found in integrated process plants. Troubleshooting guides are provided for: pumps, fans, blowers, fluidizers, conveyors, compressors, mixers, agitators, dust collectors, process rolls, gearboxes/reducers, steam traps, inverters, control valves, seals, and packing.

19

PUMPS

Design, installation, and operation are the dominant factors that affect a pump's mode of failure. This chapter identifies common modes of failure for centrifugal and positive-displacement pumps.

CENTRIFUGAL

Centrifugal pumps are especially sensitive to variations in liquid condition (i.e., viscosity, specific gravity, and temperature); suction variations, such as pressure and availability of a continuous volume of fluid; and variations in demand. Table 19–1 lists common failure modes for centrifugal pumps and their causes.

Mechanical failure may occur for a number of reasons. Some failures are induced by cavitation, hydraulic instability, or other system-related problems. Others are the direct result of improper maintenance. Maintenance-related problems include improper lubrication, misalignment, imbalance, seal leakage, and a variety of other situations that periodically affect machine reliability.

Cavitation

Cavitation in a centrifugal pump, which has a significant, negative effect on performance, is the most common failure mode. Cavitation not only degrades a pump's performance but also greatly accelerates the wear on its internal components.

Causes

Three causes of cavitation in centrifugal pumps are change of phase, entrained air or gas, and turbulent flow.

Table 19–1 Common Failure Modes of Centrifugal Pumps

THE CAUSES	Insufficient Discharge Pressure	Intermittent Operation	Insufficient Capacity	No Liquid Delivery	High Bearing Temperatures	Short Bearing Life	Short Mechanical Seal Life	High Vibration	High Noise Levels	Power Demand Excessive	Motor Trips	Elevated Motor Temperature	Elevated Liquid Temperature
Bent Shaft					●	●	●	●		●			
Casing Distorted from Excessive Pipe Strain					●	●	●	●		●		●	
Cavitation	●	●	●	●	●		●	●	●				●
Clogged Impeller	●		●	●				●		●			
Driver Imbalance						●	●	●					
Electrical Problems (Driver)					●	●	●	●			●	●	●
Entrained Air (Suction or Seal Leaks)	●	●	●					●	●			●	
Hydraulic Instability					●	●	●	●	●				
Impeller Installed Backward (Double-Suction Only)	●		●							●			
Improper Mechanical Seal							●						
Inlet Strainer Partially Clogged	●		●					●	●				●
Insufficient Flow Through Pump													●
Insufficient Suction Pressure (NPSH)	●	●	●	●				●	●				
Insufficient Suction Volume	●	●	●	●	●			●	●				●
Internal Wear	●		●					●		●			
Leakage in Piping, Valves, Vessels	●		●	●									
Mechanical Defects, Worn, Rusted, Defective Bearings					●		●			●			
Misalignment					●	●	●	●		●		●	
Misalignment (Pump and Driver)								●		●	●		●
Mismatched Pumps In Series	●		●			●		●		●			
Noncondensables in Liquid	●	●	●					●	●			●	
Obstructions in Lines or Pump Housing	●		●	●				●				●	●
Rotor Imbalance					●	●	●	●					
Specific Gravity Too High	●									●		●	
Speed Too High										●	●		
Speed Too Low	●		●	●								●	
Total System Head Higher Than Design	●	●	●	●	●		●					●	●
Total System Head Lower Than Design						●	●	●	●	●			●
Unsuitable Pumps In Parallel Operation	●		●	●	●			●	●		●		●
Viscosity Too High	●		●							●		●	
Wrong Rotation	●			●						●		●	

Source: Integrated Systems, Inc.

Change of Phase The formation or collapse of vapor bubbles in either the suction piping or inside the pump is one cause of cavitation. This failure mode normally occurs in applications, such as boiler feed, where the incoming liquid is at a temperature near its saturation point. In this situation, a slight change in suction pressure can cause the liquid to flash into its gaseous state. In the boiler-feed example, the water flashes into steam. The reverse process also can occur. A slight increase in suction pressure can force the entrained vapor to change phase to a liquid.

Cavitation due to phase change seriously damages the pump's internal components. Visual evidence of operation with phase-change cavitation is an impeller surface finish like an orange peel. Prolonged operation causes small pits or holes on both the impeller shroud and vanes.

Entrained Air or Gas Pumps are designed to handle gas-free liquids. If a centrifugal pump's suction supply contains any appreciable quantity of gas, the pump will cavitate. In the example of cavitation due to entrainment, the liquid is reasonably stable, unlike with the change of phase described in the preceding section. Nevertheless, the entrained gas has a negative effect on pump performance. While this form of cavitation does not seriously affect the pump's internal components, it severely restricts its output and efficiency.

The primary causes of cavitation due to entrained gas include two-phase suction supply, inadequate available net positive suction head ($NPSH_A$), and leakage in the suction-supply system. In some applications, the incoming liquid may contain moderate to high concentrations of air or gas. This may result from aeration or mixing the liquid prior to reaching the pump or inadequate liquid levels in the supply reservoir. Regardless of the reason, the pump is forced to handle two-phase flow, which was not intended in its design.

Turbulent Flow The effects of turbulent flow (not a true form of cavitation) on pump performance are almost identical to those described for entrained air or gas in the preceding section. Pumps are not designed to handle incoming liquids that have no stable, laminar flow pattern. Therefore, if the flow is unstable, or turbulent, the symptoms are the same as for cavitation.

Symptoms

Noise (e.g., like a can of marbles being shaken) is one indication that a centrifugal pump is cavitating. Other indications are fluctuations of the pressure gauges, flow rate, and motor current, as well as changes in the vibration profile.

Solutions

Several design or operational changes may be necessary to stop centrifugal-pump cavitation. Increasing the available net positive suction head ($NPSH_A$) above that required (NPSHR) is one way to stop it. The NPSH required to prevent cavitation is determined through testing by the pump manufacturer. It depends on several factors, including type of impeller inlet, impeller design, impeller rotational speed, pump flow rate, and the type of liquid being pumped. The manufacturer typically supplies curves of $NPSH_R$ as a function of flow rate for a particular liquid (usually water) in the pump's manual.

One way to increase the $NPSH_A$ is to increase the pump's suction pressure. If a pump is fed from an enclosed tank, suction pressure can be increased by either raising the level of the liquid in the tank or increasing the pressure in the gas space above the liquid.

The $NPSH_A$ also can be increased by decreasing the temperature of the liquid being pumped. This decreases the saturation pressure, which increases the $NPSH_A$.

If the head losses in the suction piping can be reduced, the $NPSH_A$ will be increased. Methods for reducing head losses include increasing the pipe diameter; reducing the number of elbows, valves, and fittings in the pipe; and decreasing the pipe length.

It also may be possible to stop cavitation by reducing the pump's $NPSH_R$, which is not a constant for a given pump under all conditions. Typically, the $NPSH_R$ increases significantly as the pump's flow rate increases. Therefore, reducing the flow rate by throttling a discharge valve decreases NPSHR. In addition to flow rate, $NPSH_R$ depends on pump speed. The faster the pump's impeller rotates, the greater is the $NPSH_R$. Therefore, if the speed of a variable-speed centrifugal pump is reduced, the $NPSH_R$ of the pump is decreased.

Variations in the Total System Head

Centrifugal-pump performance follows its hydraulic curve (i.e., head versus flow rate). Therefore, any variation in the total back pressure of the system causes a change in the pump's flow or output. Because pumps are designed to operate at their best efficiency point (BEP), they become more and more unstable as they are forced to operate at any other point because of changes in total system pressure, or head (TSH). This instability has a direct impact on centrifugal-pump performance, reliability, operating costs, and required maintenance.

Symptoms

The symptoms of failure due to variations in TSH include changes in motor speed and flow rate.

Motor Speed The brake horsepower of the motor that drives a pump is load dependent. As the pump's operating point deviates from BEP, the amount of horsepower required also changes. This causes a change in the pump's rotating speed, which either increases or decreases depending on the amount of work that the pump must perform.

Flow Rate The volume of liquid delivered by the pump varies with changes in TSH. An increase in the total system back pressure results in a decreased flow, while a back pressure reduction increases the pump's output.

Solutions

The best solution to problems caused by TSH variations is to prevent the variations. While it is not possible to completely eliminate them, the operating practices for centrifugal pumps should limit operation to an acceptable range of system demand for flow and pressure. If system demand exceeds the pump's capabilities, it may be necessary to change the pump, the system requirements, or both. In many applications, the pump is either too small or too large. In these instances, it is necessary to replace the pump with one that is properly sized.

For the application where the TSH is too low and the pump is operating in runout condition (i.e., maximum flow and minimum discharge pressure), the system demand can be corrected by restricting the discharge flow of the pump. This approach, called *false head*, changes the system's head by partially closing a discharge valve to increase the back pressure on the pump. Because the pump must follow its hydraulic curve, this forces the pump's performance back toward its BEP.

When the TSH is too great, there are two options: replace the pump or lower the system's back pressure by eliminating line resistance due to elbows, extra valves, and the like.

POSITIVE DISPLACEMENT

Positive-displacement pumps are more tolerant to variations in system demands and pressures than centrifugal pumps. However, they still are subject to a variety of common failure modes caused directly or indirectly by the process.

Rotary Type

Rotary-type, positive-displacement pumps share many failure modes with centrifugal pumps. Both types of pumps are subject to process-induced failure caused by demands that exceed the pump's capabilities. Process-induced failure also is caused by operating methods that either result in radical changes in their operating envelope or instability in the process system.

Table 19–2 lists common failure modes for rotary-type, positive-displacement pumps. The most common failure modes of these pumps generally are attributed to problems with the suction supply. The pumps must have a constant volume of clean liquid to function properly.

Reciprocating

Table 19–3 lists the common failure modes for reciprocating-type, positive-displacement pumps. Reciprocating pumps generally can withstand more abuse and variations in system demand than any other type. However, they must have a consistent supply of relatively clean liquid to function properly.

The weak links in the reciprocating pump's design are the inlet and discharge valves used to control pumping action. These valves are the most frequent source of failure. In most cases, valve failure is due to fatigue. The only positive way to prevent or minimize

Table 19–2 Common Failure Modes of Rotary-Type, Positive-Displacement Pumps

THE CAUSES	No Liquid Delivery	Insufficient Discharge Pressure	Insufficient Capacity	Starts, but Loses Prime	Excessive Wear	Excessive Heat	Excessive Vibration and Noise	Excessive Power Demand	Motor Trips	Elevated Motor Temperature	Elevated Liquid Temperature
Air Leakage Into Suction Piping or Shaft Seal		•	•				•			•	
Excessive Discharge Pressure			•		•		•	•	•		•
Excessive Suction Liquid Temperatures			•	•							
Insufficient Liquid Supply	•	•	•	•	•		•		•		
Internal Component Wear	•	•	•				•				
Liquid More Viscous Than Design								•	•	•	•
Liquid Vaporizing in Suction Line		•	•	•			•				•
Misaligned Coupling, Belt Drive, Chain Drive					•	•	•	•		•	
Motor or Driver Failure	•										
Pipe Strain on Pump Casing					•	•	•	•		•	
Pump Running Dry	•	•			•	•	•				
Relief Valve Stuck Open or Set Wrong		•	•								
Rotating Element Binding					•	•	•	•	•	•	
Solids or Dirt in Liquid					•						
Speed Too Low		•	•							•	
Suction Filter or Strainer Clogged	•	•	•				•			•	
Suction Piping Not Immersed in Liquid	•	•		•							
Wrong Direction of Rotation	•	•								•	

Source: Integrated Systems, Inc.

such failure is to ensure that proper maintenance is performed regularly on these components. It is important to follow the manufacturer's recommendations for valve maintenance and replacement.

Because of the close tolerances between the pistons and the cylinder walls, reciprocating pumps cannot tolerate contaminated liquid in their suction-supply system. Many of the failure modes associated with this type of pump are caused by contamination (e.g., dirt, grit, and other solids) that enter the suction side of the pump. This problem can be prevented by the use of well-maintained inlet strainers or filters.

Table 19–3 Common Failure Modes of Reciprocating Positive-Displacement Pumps

THE CAUSES	No Liquid Delivery	Insufficient Capacity	Short Packing Life	Excessive Wear Liquid End	Excessive Wear Power End	Excessive Heat Power End	Excessive Vibration and Noise	Persistent Knocking	Motor Trips
Abrasives or Corrosives in Liquid			●	●					
Broken Valve Springs		●		●			●		
Cylinders Not Filling		●	●	●			●		
Drive-Train Problems							●		●
Excessive Suction Lift	●	●							
Gear Drive Problem							●	●	●
Improper Packing Selection			●						
Inadequate Lubrication						●	●		●
Liquid Entry into Power End of Pump						●			
Loose Cross-Head Pin or Crank Pin								●	
Loose Piston or Rod								●	
Low Volumetric Efficiency		●	●						
Misalignment of Rod or Packing			●						●
Noncondensables (Air) in Liquid	●	●	●				●		●
Not Enough Suction Pressure	●	●							
Obstructions in Lines	●						●		●
One or More Cylinders Not Operating		●							
Other Mechanical Problems: Wear, Rusted, etc.					●	●	●	●	
Overloading						●			●
Pump Speed Incorrect		●				●			
Pump Valve(s) Stuck Open		●							
Relief or Bypass Valve(s) Leaking		●							
Scored Rod or Plunger		●							●
Supply Tank Empty	●								
Worn Cross-Head or Guides			●			●			
Worn Valves, Seats, Liners, Rods, or Plungers	●	●		●					

Source: Integrated Systems, Inc.

20

FANS, BLOWERS, AND FLUIDIZERS

Tables 20–1 and 20–2 list the common failure modes for fans, blowers, and fluidizers. Typical problems with these devices include output below rating, vibration and noise, and overloaded driver bearings.

CENTRIFUGAL FANS

Centrifugal fans are extremely sensitive to variations in either suction or discharge conditions. In addition to variations in ambient conditions (i.e., temperature, humidity, etc.), control variables can have a direct effect on fan performance and reliability.

Most problems that limit fan performance and reliability are caused, either directly or indirectly, by improper application, installation, operation, or maintenance. However, the majority are caused by misapplication or poor operating practices. Table 20–1 lists failure modes of centrifugal fans and their causes. Some of the more common failures are aerodynamic instability, plate-out, speed changes, and lateral flexibility.

Aerodynamic Instability

Generally, the control range of centrifugal fans is about 15 percent above and below its BEP. Fans operated outside this range tend to become progressively more unstable, which causes the fan's rotor assembly and shaft to deflect from their true centerline. This deflection increases the vibration energy of the fan and accelerates the wear on bearings and other drive-train components.

Plate Out

Dirt, moisture, and other contaminates tend to adhere to the fan's rotating element. This buildup, called *plate out*, increases the mass of the rotor assembly and decreases

Table 20–1 Common Failure Modes of Centrifugal Fans

THE CAUSES	Insufficient Discharge Pressure	Intermittent Operation	Insufficient Capacity	Overheated Bearings	Short Bearing Life	Overload on Driver	High Vibration	High Noise Levels	Power Demand Excessive	Motor Trips
Abnormal End Thrust				●			●			
Aerodynamic Instability		●	●	●	●		●	●		
Air Leaks in System	●	●	●							
Bearings Improperly Lubricated						●	●	●		●
Bent Shaft				●	●	●	●		●	
Broken or Loose Bolts or Setscrews				●			●			
Damaged Motor							●			
Damaged Wheel	●		●	●						
Dampers or Variable-Inlet Not Properly Adjusted	●		●							
Dirt in Bearings				●			●			
Excessive Belt Tension				●			●			●
External Radiated Heat				●						
Fan Delivering More Than Rated Capacity						●	●			
Fan Wheel or Driver Imbalanced				●			●			
Foreign Material in Fan Causing Imbalance (Plate-out)				●			●	●		
Incorrect Direction of Rotation	●		●			●	●			
Insufficient Belt Tension							●	●		
Loose Dampers or Variable-Inlet Vanes							●			
Misaligment of Bearings, Coupling, Wheel, or Belts				●		●	●	●	●	
Motor Improperly Wired						●	●	●		●
Packing Too Tight or Defective Stuffing Box						●	●		●	●
Poor Fan Inlet or Outlet Conditions	●		●							
Specific Gravity or Density Above Design						●	●		●	
Speed Too High		●		●	●	●	●			●
Speed Too Low	●	●	●					●		●
Too Much Grease in Ball Bearings				●						
Total System Head Greater Than Design	●		●	●		●			●	
Total System Head Less Than Design		●					●			●
Unstable Foundation		●		●			●	●		
Vibration Transmitted to Fan from Outside Sources				●			●	●		
Wheel Binding on Fan Housing				●		●	●	●		●
Wheel Mounted Backward on Shaft	●		●							
Worn Bearings							●	●		
Worn Coupling							●			
120-Cycle Magnetic Hum							●	●		

Source: Integrated Systems, Inc.

its critical speed, the point where the phenomenon referred to as *resonance* occurs. This occurs because the additional mass affects the rotor's natural frequency. Even if the fan's speed does not change, the change in natural frequency may cause its critical speed (note that machines may have more than one) to coincide with the actual rotor speed. If this occurs, the fan will resonate, or experience severe vibration, and may fail catastrophically. The symptoms of plate out often are confused with those of mechanical imbalance because both dramatically increase the vibration associated with the fan's running speed.

The problem of plate out can be resolved by regularly cleaning the fan's rotating element and internal components. Removal of buildup lowers the rotor's mass and returns its natural frequency to the initial, or design, point. In extremely dirty or dusty environments, it may be advisable to install an automatic cleaning system that uses high-pressure air or water to periodically remove any buildup that forms.

Speed Changes

In applications where a measurable fan-speed change can occur (i.e., V-belt or variable-speed drives), care must be taken to ensure that the selected speed does not coincide with any of the fan's critical speeds. For general-purpose fans, the actual running speed is designed to be between 10 and 15 percent below the first critical speed of the rotating element. If the sheave ratio of a V-belt drive or the actual running speed is increased above the design value, it may coincide with a critical speed.

Some fans are designed to operate between critical speeds. In these applications, the fan must make a transition through the first critical speed to reach its operating speed. Such transitions must be made as quickly as possible to prevent damage. If the fan's speed remains at or near the critical speed for any extended period of time, serious damage can occur.

Lateral Flexibility

By design, the structural support of most general-purpose fans lacks the mass and rigidity needed to prevent flexing of the fan's housing and rotating assembly. This problem is more pronounced in the horizontal plane but also present in the vertical direction. If support-structure flexing is found to be the root cause or a major contributing factor to the problem, it can be corrected by increasing the stiffness or mass of the structure. However, do not fill the structure with concrete. As it dries, concrete pulls away from the structure and does little to improve its rigidity.

BLOWERS OR POSITIVE-DISPLACEMENT FANS

Blowers, or positive-displacement fans, have the same common failure modes as rotary pumps and compressors. Table 20–2 (also see Tables 19–2 and 22–2), lists the failure modes that most often affect blowers and fluidizers. In particular, blower fail-

Table 20–2 Common Failure Modes of Blowers and Fluidizers

THE CAUSES	No Air/Gas Delivery	Insufficient Discharge Pressure	Insufficient Capacity	Excessive Wear	Excessive Heat	Excessive Vibration and Noise	Excessive Power Demand	Motor Trips	Elevated Motor Temperature	Elevated Air/Gas Temperature
Air Leakage into Suction Piping or Shaft Seal		•	•			•				
Coupling Misaligned				•	•	•	•		•	
Excessive Discharge Pressure			•	•		•	•	•		•
Excessive Inlet Temperature/Moisture			•							
Insufficient Suction Air/Gas Supply		•	•	•		•			•	
Internal Component Wear	•	•	•							
Motor or Driver Failure	•									
Pipe Strain on Blower Casing				•	•	•	•		•	
Relief Valve Stuck Open or Set Wrong		•	•							
Rotating Element Binding				•	•	•	•	•	•	
Solids or Dirt in Inlet Air/Gas Supply				•						
Speed Too Low		•	•						•	
Suction Filter or Strainer Clogged	•	•	•			•			•	
Wrong Direction of Rotation	•	•							•	

Source: Integrated Systems, Inc.

ures occur due to process instability, caused by start/stop operation and demand variations, and mechanical failures due to close tolerances.

Process Instability

Blowers are very sensitive to variations in their operating envelope. As little as a 1 psig change in downstream pressure can cause the blower to become extremely unstable. The probability of catastrophic failure or severe damage to blower components increases in direct proportion to the amount and speed of the variation in demand or downstream pressure.

Start/Stop Operation

The transients caused by frequent start/stop operation also have a negative effect on blower reliability. Conversely, blowers that operate constantly in a stable environment rarely exhibit problems. The major reason is the severe axial thrusting caused by the frequent variations in suction or discharge pressure caused by the start/stop operation.

Demand Variations

Variations in pressure and volume demands have a serious impact on blower reliability. Since blowers are positive-displacement devices, they generate a constant volume and a variable pressure that is dependent on the downstream system's back pressure. If demand decreases, the blower's discharge pressure continues to increase until (1) a downstream component fails and reduces the back pressure or (2) the brake horsepower required to drive the blower is greater than the motor's locked rotor rating. Either of these result in failure of the blower system. The former may result in a reportable release, while the latter will cause the motor to trip or burn out.

Frequent variations in demand greatly accelerate the wear on the thrust bearings in the blower. This can be directly attributed to the constant, instantaneous axial thrusting caused by variations in the discharge pressure required by the downstream system.

Mechanical Failures

Because of the extremely close clearances that must exist within the blower, the potential for serious mechanical damage or catastrophic failure is higher than with other rotating machinery. The primary failure points include thrust bearing, timing gears, and rotor assemblies.

In many cases, these mechanical failures are caused by the instability discussed in the preceding sections, but poor maintenance practices are another major cause. See the troubleshooting guide Table 22–2 for rotary-type, positive-displacement compressors for more information.

21

CONVEYORS

Conveyor failure modes vary, depending on the type of system. Two common types of conveyor systems used in chemical plants are pneumatic and chain-type mechanical.

PNEUMATIC

Table 21–1 lists common failure modes associated with pneumatic-conveyor systems. However, most common problems can be attributed to either conveyor piping plugging or problems with the prime mover (i.e., fan or fluidizer). For a centrifugal fan troubleshooting guide, refer to Table 20–1. For fluidizer and blower guides, refer to Table 20–2.

CHAIN-TYPE MECHANICAL

The Hefler-type chain conveyor is a very common type of mechanical conveyor used in integrated chemical plants. Table 21–2 provides the more common failure modes of this type of conveyor. Most of the failure modes defined in the table can be attributed directly to operating practices, changes in incoming product quality (i.e., density or contamination), or maintenance practices.

Table 21–1 Common Failure Modes of Pneumatic Conveyors

THE CAUSES	Fails to Deliver Rated Capacity	Output Exceeds Rated Capacity	Frequent Fan/Blower Motor Trips	Product Contamination	Frequent System Blockage	Fan/Blower Failures	Fan/Blower Bearing Failures
Aerodynamic Imbalance			●			●	●
Blockage Caused By Compaction of Product	●		●			●	
Contamination in Incoming Product				●			
Excessive Moisture in Product/Piping	●		●	●	●	●	
Fan/Blower Too Small	●		●			●	
Foreign Object Blocking Piping	●		●		●		
Improper Lubrication						●	●
Mechanical Imbalance						●	●
Misalignment						●	●
Piping Configuration Unsuitable	●		●		●		
Piping Leakage	●			●			
Product Compaction During Downtime/Stoppage	●		●		●		
Product Density Too Great	●		●			●	
Product Density Too Low		●					
Rotor Binding or Contacting			●			●	●
Startup Torque Too Great			●				

Table 21–2 Common Failure Modes of Hefler-Type Chain Conveyors

THE CAUSES	Fails To Deliver Rated Capacity	Frequent Drive Motor Trips	Conveyor Blockage	Abnormal Wear on Drive Gears	Excessive Shear Pin Breakage	Excessive Bearing Failures/Wear	Motor Overheats	Excessive Noise
Blockage of Conveyor Ductwork	●	●					●	
Chain Misaligned			●		●	●	●	●
Conveyor Chain Binding on Ductwork								●
Conveyor Not Emptied Before Shut-down		●	●		●			
Conveyor Overfilled When Idle		●	●		●			
Excesssive Looseness on Drive Chains	●							
Excessive Moisture in Product	●	●	●					
Foreign Object Obstructing Chain	●	●				●	●	
Gear Set Center-to-Center Distance Incorrect				●				●
Gears Misaligned				●		●	●	●
Lack of Lubrication				●		●	●	●
Motor Speed Control Damaged or Not Calibrated	●							
Product Density Too High	●	●			●		●	
Too Much Volume/Load	●	●					●	

THE PROBLEM

22

COMPRESSORS

Compressors can be divided into three classifications: centrifugal, rotary, and reciprocating. This section identifies the common failure modes for each.

CENTRIFUGAL

The operating dynamics of centrifugal compressors are the same as for other centrifugal machine trains. The dominant forces and vibration profiles typically are identical to pumps or fans. However, the effects of variable load and other process variables (e.g., temperatures, inlet/discharge pressure) are more pronounced than in other rotating machines. Table 22–1 identifies the common failure modes for centrifugal compressors.

Aerodynamic instability is the most common failure mode for centrifugal compressors. Variable demand and restrictions of the inlet airflow are common sources of this instability. Even slight variations can cause dramatic changes in the operating stability of the compressor.

The operating life also can be affected by entrained liquids and solids. When dirty air must be handled, open-type impellers should be used. An open design provides the ability to handle a moderate amount of dirt or other solids in the inlet-air supply. However, inlet filters are recommended for all applications and controlled liquid injection for cleaning and cooling should be considered during the design process.

ROTARY, POSITIVE DISPLACEMENT

Table 22–2 lists the common failure modes of rotary-type, positive-displacement compressors. This type of compressor can be grouped into two types: sliding vane and rotary screw.

Table 22–1 Common Failure Modes of Centrifugal Compressors

THE CAUSES	THE PROBLEM								
	Excessive Vibration	Compressor Surges	Loss of Discharge Pressure	Low Lube Oil Pressure	Excessive Bearing Oil Drain Temp.	Units Do Not Stay in Alignment	Persistent Unloading	Water in Lube Oil	Motor Trips
Bearing Lube Oil Orifice Missing or Plugged				●					
Bent Rotor (Caused by Uneven Heating and Cooling)	●						●		
Buildup of Deposits on Diffuser		●							
Buildup of Deposits on Rotor	●	●							
Change in System Resistance		●							●
Clogged Oil Strainer/Filter				●					
Compressor Not up to Speed			●						
Condensate in Oil Reservoir								●	
Damaged Rotor	●								
Dry Gear Coupling	●								
Excessive Bearing Clearance	●								
Excessive Inlet Temperature			●						
Failure of Both Main and Auxiliary Oil Pumps				●					
Faulty Temperature Gauge or Switch				●	●				●
Improperly Assembled Parts	●						●		●
Incorrect Pressure Control Valve Setting				●					
Insufficient Flow		●							
Leak In Discharge Piping			●						
Leak In Lube Oil Cooler Tubes or Tube Sheet								●	
Leak in Oil Pump Suction Piping				●					
Liquid "Slugging"	●						●		
Loose or Broken Bolting	●								
Loose Rotor Parts	●								
Oil Leakage				●					
Oil Pump Suction Plugged				●					
Oil Reservoir Low Level				●					
Operating at Low Speed w/o Auxiliary Oil Pump				●					
Operating in Critical Speed Range	●								
Operating in Surge Region	●								
Piping Strain	●					●	●	●	●
Poor Oil Condition					●				
Relief Valve Improperly Set or Stuck Open				●					
Rotor Imbalance	●						●		
Rough Rotor Shaft Journal Surface					●		●		●
Shaft Misalignment	●					●			
Sympathetic Vibration	●						●	●	
Vibration					●				
Warped Foundation or Baseplate							●		●
Wiped or Damaged Bearings					●				●
Worn or Damaged Coupling	●								

Table 22–2 Common Failure Modes of Rotary, Positive-Displacement Compressors

THE CAUSES	No Air/Gas Delivery	Insufficient Discharge Pressure	Insufficient Capacity	Excessive Wear	Excessive Heat	Excessive Vibration and Noise	Excessive Power Demand	Motor Trips	Elevated Motor Temperature	Elevated Air/Gas Temperature
Air Leakage Into Suction Piping or Shaft Seal		●	●			●				
Coupling Misaligned				●	●	●	●		●	
Excessive Discharge Pressure			●	●		●	●	●		●
Excessive Inlet Temperature/Moisture			●							
Insufficient Suction Air/Gas Supply		●	●	●		●		●		
Internal Component Wear	●	●	●							
Motor or Driver Failure	●									
Pipe Strain on Compressor Casing				●	●	●	●		●	
Relief Valve Stuck Open or Set Wrong	●	●								
Rotating Element Binding				●	●	●	●	●	●	
Solids or Dirt in Inlet Air/Gas Supply				●						
Speed Too Low		●	●					●		
Suction Filter or Strainer Clogged	●	●	●			●			●	
Wrong Direction of Rotation	●	●							●	

(Column group heading: **THE PROBLEM**)

Sliding Vane

Sliding-vane compressors have the same failure modes as vane-type pumps. The dominant components in their vibration profile are running speed, vane-pass frequency, and bearing-rotation frequency. In normal operation, the dominant energy is at the shaft's running speed. The other frequency components are at much lower energy levels. Common failures of this type of compressor occur with shaft seals, vanes, and bearings.

Shaft Seals

Leakage through the shaft's seals should be checked visually once a week or as part of every data-acquisition route. Leakage may not be apparent from the outside of the gland. If the fluid is removed through a vent, the discharge should be configured for easy inspection. Generally, more leakage than normal is the signal to replace a seal. Under good conditions, the seals have a normal life of 10,000 to 15,000 hours and should be replaced routinely when this service life has been reached.

Vanes

Vanes wear continuously on their outer edges and, to some degree, on the faces that slide in and out of the slots. The vane material is affected somewhat by prolonged heat, which causes gradual deterioration. The typical life expectancy of vanes in 100 psig service is about 16,000 hours of operation. For low-pressure applications, life-times may reach 32,000 hours.

Replacing vanes before they break is extremely important. Breakage during operation can severely damage the compressor, which requires a complete overhaul and realignment of heads and clearances.

Bearings

In normal service, bearings have a relatively long life. Replacement after about six years of operation generally is recommended. Bearing defects usually are displayed in the same manner in a vibration profile as for any rotating machine train. Inner and outer race defects are the dominant failure modes, but roller spin also may contribute to the failure.

Rotary Screw

The most common reason for compressor failure or component damage is process instability. Rotary-screw compressors are designed to deliver a constant volume and pressure of air or gas. These units are extremely susceptible to any change in either inlet or discharge conditions. A slight variation in pressure, temperature, or volume can result in instantaneous failure. The following are used as indices of instability and potential problems: rotor mesh, axial movement, thrust bearings, and gear mesh.

Rotor Mesh

In normal operation, the vibration energy generated by male and female rotor meshing is very low. As the process becomes unstable, the energy due to the rotor-meshing frequency increases, with both the amplitude of the meshing frequency and the width of the peak increasing. In addition, the noise floor surrounding the meshing frequency becomes more pronounced. This white noise is similar to that observed in a cavitating pump or unstable fan.

Axial Movement

The normal tendency of the rotors and helical timing gears is to generate axial shaft movement, or thrusting. However, the extremely tight clearances between the male and female rotors do not tolerate any excessive axial movement, and therefore, axial movement should be a primary monitoring parameter. Axial measurements are needed from both rotor assemblies. If there is any increase in the vibration amplitude of these measurements, it is highly probable that the compressor will fail.

Thrust Bearings

While process instability can affect both the fixed and floating bearings, the thrust bearing is more likely to show early degradation as a result of process instability or abnormal compressor dynamics. Therefore, these bearings should be monitored closely, and any degradation or hint of excessive axial clearance should be corrected immediately.

Gear Mesh

The gear mesh vibration profile also provides an indication of prolonged compressor instability. Deflection of the rotor shafts changes the wear pattern on the helical gear sets. This change in pattern increases the backlash in the gear mesh, results in higher vibration levels, and increases thrusting.

RECIPROCATING, POSITIVE DISPLACEMENT

Reciprocating compressors have a history of chronic failures that include valves, lubrication system, pulsation, and imbalance. Table 22–3 identifies common failure modes and causes for this type of compressor.

Like all reciprocating machines, reciprocating compressors normally generate higher levels of vibration than centrifugal machines. In part, the increased level of vibration is due

Table 22–3a Common Failure Modes of Reciprocating, Positive-Displacement Compressors

THE CAUSES	Air Discharge Temperature Above Normal	Carbonaceous Deposits Abnormal	Compressor Fails to Start	Compressor Fails to Unload	Compressor Noisy or Knocks	Compressor Parts Overheat	Crankcase Oil Pressure Low	Crankcase Water Accumulation	Delivery Less Than Rated Capacity	Discharge Pressure Below Normal	Excessive Compressor Vibration	Intercooler Pressure Above Normal	Intercooler Pressure Below Normal	Intercooler Safety Valve Pops	Motor Overheating	Oil Pumping Excessive (Single-Acting Compressor)	Operating Cycle Abnormally Long	Outlet Water Temperature Above Normal	Piston Ring, Piston, Cylinder Wear Excessive	Piston Rod or Packing Wear Excessive	Receiver Pressure Above Normal	Receiver Safety Valve Pops	Starts Too Often	Valve Wear and Breakage Normal
Air Discharge Temperature Too High		•															•							
Air Filter Defective		•																	•	•				•
Air Flow to Fan Blocked	•	•			•																			
Air Leak into Pump Suction						•																		
Ambient Temperature Too High	•	•			•								•											
Assembly Incorrect																								•
Bearings Need Adjustment or Renewal				•	•	•							•											
Belts Slipping				•					•	•														
Belts Too Tight			•		•								•											
Centrifugal Pilot Valve Leaks																•								
Check or Discharge Valve Defective					•																			
Control Air Filter, Strainer Clogged				•																				
Control Air Line Clogged																					•			
Control Air Pipe Leaks																					•	•		
Crankcase Oil Pressure Too High															•									
Crankshaft End Play Too Great				•																				
Cylinder, Head, Cooler Dirty	•	•																						
Cylinder, Head, Intercooler Dirty					•													•						
Cylinder (Piston) Worn or Scored	•	•		•	•		•	•	•	●H	●L	●H	●L	•	•				●H	●H				
Detergent Oil Being Used (3)								•																
Demand Too Steady (2)																							•	
Dirt, Rust Entering Cylinder		•																	•	•				•

Table 22–3b Common Failure Modes of Reciprocating, Positive-Displacement Compressors

THE PROBLEM

THE CAUSES	Air Discharge Temperature Above Normal	Carbonaceous Deposits Abnormal	Compressor Fails to Start	Compressor Fails to Unload	Compressor Noisy or Knocks	Compressor Parts Overheat	Crankcase Oil Pressure Low	Crankcase Water Accumulation	Delivery Less Than Rated Capacity	Discharge Pressure Below Normal	Excessive Compressor Vibration	Intercooler Pressure Above Normal	Intercooler Pressure Below Normal	Intercooler Safety Valve Pops	Motor Overheating	Oil Pumping Excessive (Single-Acting Compressor)	Operating Cycle Abnormally Long	Outlet Water Temperature Above Normal	Piston Ring, Piston, Cylinder Wear Excessive	Piston Rod or Packing Wear Excessive	Receiver Pressure Above Normal	Receiver Safety Valve Pops	Starts Too Often	Valve Wear and Breakage Normal
Discharge Line Restricted	•														•									
Discharge Pressure Above Rating	•	•			•	•			•			•	•		•		•	•	•	•	•	•		
Electrical Conditions Wrong			•												•									
Excessive Number of Starts															•									
Excitation Inadequate			•												•									
Foundation Bolts Loose					•						•													
Foundation Too Small											•													
Foundation Uneven–Unit Rocks					•						•													
Fuses Blown			•																					
Gaskets Leak	•	•			•	•			•	•		•H	•L	•H	•L		•		•H	•H				
Gauge Defective							•			•		•	•								•			
Gear Pump Worn/Defective							•																	
Grout, Improperly Placed											•													
Intake Filter Clogged	•				•	•			•	•			•				•	•	•					
Intake Pipe Restricted, Too Small, Too Long	•				•	•			•	•			•				•	•	•					
Intercooler, Drain More Often								•																
Intercooler Leaks														•										
Intercooler Passages Clogged												•		•										
Intercooler Pressure Too High																	•							
Intercooler Vibrating					•																			
Leveling Wedges Left Under Compressor											•													
Liquid carry-over					•			•											•	•				•

Table 22–3c Common Failure Modes of Reciprocating, Positive-Displacement Compressors

THE PROBLEM

THE CAUSES	Air Discharge Temperature Above Normal	Carbonaceous Deposits Abnormal	Compressor Fails to Start	Compressor Fails to Unload	Compressor Noisy or Knocks	Compressor Parts Overheat	Crankcase Oil Pressure Low	Crankcase Water Accumulation	Delivery Less Than Rated Capacity	Discharge Pressure Below Normal	Excessive Compressor Vibration	Intercooler Pressure Above Normal	Intercooler Pressure Below Normal	Intercooler Safety Valve Pops	Motor Overheating	Oil Pumping Excessive (Single-Acting Compressor)	Operating Cycle Abnormally Long	Outlet Water Temperature Above Normal	Piston Ring, Piston, Cylinder Wear Excessive	Piston Rod or Packing Wear Excessive	Receiver Pressure Above Normal	Receiver Safety Valve Pops	Starts Too Often	Valve Wear and Breakage Normal
Location Too Humid and Damp								•																
Low Oil Pressure Relay Open			•																					
Lubrication Inadequate	•				•	•									•		•		•	•				•
Motor Overload Relay Tripped			•																					
Motor Rotor Loose on Shaft					•						•													
Motor Too Small			•												•									
New Valve on Worn Seat																								•
"Off" Time Insufficient	•	•			•																			
Oil Feed Excessive		•			•															•				•
Oil Filter or Strainer Clogged							•																	
Oil Level Too High	•	•			•	•												•						
Oil Level Too Low						•	•																	
Oil Relief Valve Defective							•																	
Oil Viscosity Incorrect		•			•	•	•								•	•			•	•				•
Oil Wrong Type																•								
Packing Rings Worn, Stuck, Broken																				•				
Piping Improperly Supported											•													
Piston or Piston Nut Loose					•																			
Piston or Ring Drain Hole Clogged																•								
Piston Ring Gaps Not Staggered																•								
Piston Rings Worn, Broken, or Stuck	•	•			•	•		•	•	•		•H	•L	•H	•L	•	•		•H	•H				
Piston-to-head Clearance Too Small					•																			

Table 22–3d Common Failure Modes of Reciprocating, Positive-Displacement Compressors

THE PROBLEM

THE CAUSES	Air Discharge Temperature Above Normal	Carbonaceous Deposits Abnormal	Compressor Fails to Start	Compressor Fails to Unload	Compressor Noisy or Knocks	Compressor Parts Overheat	Crankcase Oil Pressure Low	Crankcase Water Accumulation	Delivery Less Than Rated Capacity	Discharge Pressure Below Normal	Excessive Compressor Vibration	Intercooler Pressure Above Normal	Intercooler Pressure Below Normal	Intercooler Safety Valve Pops	Motor Overheating	Oil Pumping Excessive (Single-Acting Compressor)	Operating Cycle Abnormally Long	Outlet Water Temperature Above Normal	Piston Ring, Piston, Cylinder Wear Excessive	Piston Rod or Packing Wear Excessive	Receiver Pressure Above Normal	Receiver Safety Valve Pops	Starts Too Often	Valve Wear and Breakage Normal
Pulley or Flywheel Loose					•						•													
Receiver, Drain More Often																							•	
Receiver Too Small																							•	
Regulation Piping Clogged				•																				
Resonant Pulsation (Inlet or Discharge)												•	•	•	•									•
Rod Packing Leaks	•				•	•			•	•														
Rod Packing Too Tight						•																		
Rod Scored, Pitted, Worn																				•				
Rotation Wrong	•	•	•																					
Runs Too Little *(2)*								•																
Safety Valve Defective												•	•									•		
Safety Valve Leaks	•				•				•	•		•		•										
Safety Valve Set Too Low													•									•		
Speed Demands Exceed Rating																	•							
Speed Lower Than Rating									•	•														
Speed Too High	•	•			•						•				•			•						
Springs Broken																								•
System Demand Exceeds Rating	•				•				•	•		•			•		•							
System Leakage Excessive	•				•				•	•		•		•	•		•					•		
Tank Ringing Noise					•																			
Unloader Running Time Too Long *(1)*																•								
Unloader or Control Defective	•	•	•	•	•	•			•	•	•	•	•	•	•		•			•	•	•	•	•

Table 22–3e Common Failure Modes of Reciprocating, Positive-Displacement Compressors

THE PROBLEM

THE CAUSES	Air Discharge Temperature Above Normal	Carbonaceous Deposits Abnormal	Compressor Fails to Start	Compressor Fails to Unload	Compressor Noisy or Knocks	Compressor Parts Overheat	Crankcase Oil Pressure Low	Crankcase Water Accumulation	Delivery Less Than Rated Capacity	Discharge Pressure Below Normal	Excessive Compressor Vibration	Intercooler Pressure Above Normal	Intercooler Pressure Below Normal	Intercooler Safety Valve Pops	Motor Overheating	Oil Pumping Excessive (Single-Acting Compressor)	Operating Cycle Abnormally Long	Outlet Water Temperature Above Normal	Piston Ring, Piston, Cylinder Wear Excessive	Piston Rod or Packing Wear Excessive	Receiver Pressure Above Normal	Receiver Safety Valve Pops	Starts Too Often	Valve Wear and Breakage Normal
Unloader Parts Worn or Dirty				•																				
Unloader Setting Incorrect	•	•			•	•			•	•		•	•	•	•		•			•	•	•	•	•
V-belt or Other Misalignment					•	•					•													
Valves Dirty	•	•				•						•	•											•
Valves Incorrectly Located	•	•			•	•			•	•		•H	•L	•H	•L		•		•H	•H				
Valves Not Seated in Cylinder	•	•			•	•			•	•		•H	•L	•H	•L				•H	•H				
Valves Worn or Broken	•	•			•	•			•	•		•H	•L	•H	•L		•H	•H	•H	•H				
Ventilation Poor	•	•				•									•									
Voltage Abnormally Low			•												•									
Water Inlet Temperature Too High	•	•				•			•			•						•						
Water Jacket or Cooler Dirty	•	•																						
Water Jackets or Intercooler Dirty						•						•						•						
Water Quantity Insufficient	•					•			•			•						•						
Wiring Incorrect			•																					
Worn Valve on Good Seat																								•
Wrong Oil Type		•																			•	•		
(1) Use Automatic Start/Stop Control																								
(2) Use Constant Speed Control																								
(3) Change to Non-detergent Oil																								
H (in High Pressure Cylinder)																								
L (in Low Pressure Cylinder)																								

to the impact as each piston reaches the top and bottom dead-center of its stroke. The energy levels also are influenced by the unbalanced forces generated by nonopposed pistons and looseness in the piston rods, wrist pins, and journals of the compressor. In most cases, the dominant vibration frequency is the second harmonic (2X) of the main crankshaft's rotating speed. Again, this results from the impact that occurs when each piston changes direction (i.e., two impacts occur during one complete crankshaft rotation).

Valves

Valve failure is the dominant failure mode for reciprocating compressors. Because of their high cyclic rate, which exceeds 80 million cycles per year, inlet and discharge valves tend to work harden and crack.

Lubrication System

Poor maintenance of lubrication-system components, such as filters and strainers, typically causes premature failure. Such maintenance is crucial to reciprocating compressors because they rely on the lubrication system to provide a uniform oil film between closely fitting parts (e.g., piston rings and the cylinder wall). Partial or complete failure of the lube system results in catastrophic failure of the compressor.

Pulsation

Reciprocating compressors generate pulses of compressed air or gas that are discharged into the piping that transports the air or gas to its point(s) of use. This pulsation often generates resonance in the piping system, and pulse impact (i.e., standing waves) can severely damage other machinery connected to the compressed-air system. While this behavior does not cause the compressor to fail, it must be prevented to protect other plant equipment. Note, however, that most compressed-air systems do not use pulsation dampers.

Each time the compressor discharges compressed air, the air tends to act like a compression spring. Because it rapidly expands to fill the discharge piping's available volume, the pulse of high-pressure air can cause serious damage. The pulsation wavelength, λ, from a compressor having a double-acting piston design can be determined by

$$\lambda = \frac{60a}{2n} = \frac{34,050}{n}$$

where

λ = wavelength, ft;
a = speed of sound = 1,135 ft/sec;
n = compressor speed, revolutions/min.

For a double-acting piston design, a compressor running at 1,200 rpm will generate a standing wave of 28.4 ft. In other words, a shock load equivalent to the discharge pressure will be transmitted to any piping or machine connected to the discharge piping and located within 28 ft of the compressor. Note that, for a single-acting cylinder, the wavelength will be twice as long.

Imbalance

Compressor inertial forces may have two effects on the operating dynamics of a reciprocating compressor, affecting its balance characteristics. The first is a force in the direction of the piston movement, which is displayed as impacts in a vibration profile as the piston reaches top and bottom dead-center of its stroke. The second effect is a couple, or moment, caused by an offset between the axes of two or more pistons on a common crankshaft. The interrelationship and magnitude of these two effects depend on such factors as number of cranks, longitudinal and angular arrangement, cylinder arrangement, and amount of counterbalancing possible. Two significant vibration periods result, the primary at the compressor's rotation speed (X) and the secondary at 2X.

Although the forces developed are sinusoidal, only the maximum (i.e., the amplitude) is considered in the analysis. Figure 22–1 shows relative values of the inertial forces for various compressor arrangements.

CRANK ARRANGEMENTS	FORCES		COUPLES	
	PRIMARY	SECONDARY	PRIMARY	SECONDARY
SINGLE CRANK	F' WITHOUT COUNTERWTS. $0.5F'$ WITH COUNTERWTS.	F''	NONE	NONE
TWO CRANKS AT 180° IN LINE CYLINDERS	ZERO	$2F''$	$F'D$ WITHOUT COUNTERWTS. $\frac{F'D}{2}$ WITH COUNTERWTS.	NONE
TWO CRANKS AT 180° OPPOSED CYLINDERS	ZERO	ZERO	NIL	NIL
TWO CRANKS AT 90°	1.41 F' WITHOUT COUNTERWTS. 0.707 F' WITH COUNTERWTS.	ZERO	.707 $F'D$ WITHOUT COUNTERWTS. 0.354 $F'D$ WITH COUNTERWTS.	$F''D$
TWO CYLINDERS ON ONE CRANK CYLINDERS AT 90°	F' WITHOUT COUNTERWTS. ZERO WITH COUNTERWTS.	1.41 F''	NIL	NIL
TWO CYLINDERS ON ONE CRANK OPPOSED CYLINDERS	$2F'$ WITHOUT COUNTERWTS. F' WITH COUNTERWTS.	ZERO	NONE	NIL
THREE CRANKS AT 120°	ZERO	ZERO	3.46 $F'D$ WITHOUT COUNTERWTS. 1.73 $F'D$ WITH COUNTERWTS.	3.46 $F''D$
FOUR CYLINDERS CRANKS AT 180°	ZERO	$4F''$	ZERO	ZERO
FOUR CYLINDERS CRANKS AT 90°	ZERO	ZERO	1.41 $F'D$ WITHOUT COUNTERWTS. 0.707 $F'D$ WITH COUNTERWTS.	4.0 $F''D$
SIX CYLINDERS	ZERO	ZERO	ZERO	ZERO

F' = PRIMARY INERTIA FORCE IN LBS.
$F' = .0000284\ RN^2W$
F'' = SECONDARY INERTIA FORCE IN LBS.
$F'' = \frac{R}{L} F'$
R = CRANK RADIUS, INCHES
N = R.P.M.
W = RECIPROCATING WEIGHT OF ONE CYLINDER, LBS
L = LENGTH OF CONNECTING ROD, INCHES
D = CYLINDER CENTER DISTANCE

Figure 22–1 Unbalanced inertial forces and couples for various reciprocating compressors (Gibbs 1971).

23

MIXERS AND AGITATORS

Table 23–1 identifies common failure modes and their causes for mixers and agitators. Most of the problems that affect performance and reliability are caused by improper installation or variations in the product's physical properties.

Proper installation of mixers and agitators is critical. The physical location of the vanes or propellers within the vessel is the dominant factor to consider. If the vanes are set too close to the side, corner, or bottom of the vessel, a stagnant zone will develop that causes both loss of mixing quality and premature damage to the equipment. If the vanes are set too close to the liquid level, vortexing can develop. This also will cause a loss of efficiency and accelerated component wear.

Variations in the product's physical properties, such as viscosity, will cause loss of mixing efficiency and premature wear of mixer components. Although the initial selection of the mixer or agitator may have addressed the full range of physical properties expected to be encountered, applications sometimes change. Such a change may result in the use of improper equipment for a particular application.

Table 23–1 Common Failure Modes of Mixers and Agitators

THE CAUSES	Surface Vortex Visible	Incomplete Mixing Of Product	Excessive Vibration	Excessive Wear	Motor Overheats	Excessive Power Demand	Excessive Bearing Failures
Abrasives In Product				●			
Mixer/Agitator Setting Too Close To Side or Corner		●	●	●		●	●
Mixer/Agitator Setting Too High	●	●					
Mixer/Agitator Setting Too Low		●			●		
Mixer/Agitator Shaft Too Long							●
Product Temperature Too Low		●			●	●	
Rotating Element Imbalanced or Damaged		●	●		●	●	●
Speed Too High	●		●	●			
Speed Too Low		●					
Viscosity/Specific Gravity Too High		●			●	●	
Wrong Direction Of Rotation		●			●		●

Source: Integrated Systems, Inc.

24

DUST COLLECTORS

This chapter identifies common problems and their causes for baghouse and cyclonic separator dust-collection systems.

BAGHOUSES

Table 24–1 lists the common failure modes for baghouses. This guide may be used for all such units that use fabric filter bags as the primary dust-collection media.

CYCLONIC SEPARATORS

Table 24–2 identifies the failure modes and their causes for cyclonic separators. Since cyclonic separators have no moving parts, most of the problems associated with this type of system can be attributed to variations in process parameters, such as flow rate, dust load, dust composition (i.e., density, size, etc.), and ambient conditions (i.e., temperature, humidity, etc.).

Table 24–1 Common Failure Modes of Baghouses

THE CAUSES	Continuous Release of Dust-laden Air	Intermittent Release of Dust-laden Air	Loss of Plant Air Pressure	Blow-down Ineffective	Insufficient Capacity	Excessive Differential Pressure	Fan/Blower Motor Trips	Fan Has High Vibration	Premature Bag Failures	Differential Pressure Too Low	Chronic Plugging of Bags
Bag Material Incompatible For Application									●		●
Bag Plugged						●	●	●			
Bag Torn or Improperly Installed	●							●	●	●	
Baghouse Undersized					●		●				●
Blow-down Cycle Interval Too Long						●	●				
Blow-down Cycle Time Failed or Damaged						●	●				
Blow-down Nozzles Plugged						●					
Blow-down Pilot Valve Failed To Open (Solenoid Failure)				●		●					
Dust Load Exceeds Capacity											●
Excessive Demand				●							
Fan/Blower Not Operating Properly					●						
Improper or Inadequate Lubrication								●			
Leaks In Ductwork or Baghouse	●				●						
Misalignment of Fan and Motor								●			
Moisture Content Too High											●
Not Enough Blow-down Air (Pressure and Volume)			●	●		●					
Not Enough Dust Layer on Filter Bags	●	●						●		●	
Piping/Valve Leaks			●								
Plate-out (Dust Build-up on Fan's Rotor)								●			
Plenum Cracked or Seal Defective	●		●							●	
Rotor Imbalanced								●			
Ruptured Blow-down Diaphrams			●	●		●					
Suction Ductwork Blocked or Plugged					●						

Source: Integrated Systems, Inc.

Table 24–2 Common Failure Modes of Cyclonic Separators

THE CAUSES	Continuous Release of Dust-Laden Air	Intermittent Release of Dust-Laden Air	Cyclone Plugs in Inlet Chamber	Cyclone Plugs in Dust Removal Section	Rotor-Lock Valve Fails To Turn	Excessive Differential Pressure	Differential Pressure Too Low	Rotor-Lock Valve Leaks	Fan Has High Vibration
Clearance Set Wrong								●	
Density and Size Distribution of Dust Too High				●	●	●			●
Density and Size Distribution of Dust Too Low	●	●							
Dust Load Exceeds Capacity	●	●				●			●
Excessive Moisture In Incoming Air			●						
Foreign Object Lodged In Valve					●				
Improper Drive-Train Adjustments					●				
Improper Lubrication					●				
Incoming Air Velocity Too High						●			
Incoming Air Velocity Too Low	●	●	●				●		
Internal Wear or Damage								●	
Large Contaminates in Incoming Air Stream			●		●				
Prime Mover (Fan, Blower) Malfunctioning	●	●				●	●		●
Rotor-Lock Valve Turning Too Slow	●	●		●					
Seals Damaged								●	

Source: Integrated Systems, Inc.

25

PROCESS ROLLS

Most failures that cause reliability problems with process rolls can be attributed to either improper installation or abnormal induced loads. Table 25–1 identifies the common failure modes of process rolls and their causes.

Installation problems normally result from misalignment, where the roll is not perpendicular to the travel path of the belt or transported product. If process rolls are misaligned, either vertically or horizontally, the load imparted by the belt or the carried product is not spread uniformly across the roll face or to the support bearings. As a result, both the roll face and bearings are subjected to abnormal wear and may fail prematurely.

Operating methods may cause induced loads that are outside the acceptable design limits of the roll or its support structure. Operating variables, such as belt or strip tension or tracking, may be the source of chronic reliability problems. As with misalignment, these variables apply an unequal load distribution across the roll face and bearing-support structure. These abnormal loads accelerate wear and may result in premature failure of the bearings or roll.

Table 25–1 Common Failure Modes of Process Rolls

THE CAUSES	Frequent Bearing Failures	Abnormal Roll Face Wear	Roll Neck Damage or Failure	Abnormal Product Tracking	Motor Overheats	Excessive Power Demand	High Vibration	Product Quality Poor
Defective or Damaged Roll Bearings								●
Excessive Product Tension	●	●	●	●	●	●		●
Excessive Load					●	●		
Misaligned Roll	●	●	●	●	●	●	●	●
Poor Roll Grinding Practices								●
Product Tension Too Loose								●
Product Tension/Tracking Problem		●		●				●
Roll Face Damage	●		●	●				●
Speed Coincides With Roll's Natural Frequency	●			●			●	●
Speed Coincides With Structural Natural Frequency		●		●			●	●

Source: Integrated Systems, Inc.

26

GEARBOXES OR REDUCERS

This chapter identifies common gearbox (also called *reducer*) problems and their causes. Table 26–1 lists the more common gearbox failure modes. A primary cause of failure is that, with few exceptions, gear sets are designed for operation in one direction only. Failure often is caused by inappropriate bidirectional operation of the gearbox or backward installation of the gear set. Unless specifically manufactured for bidirectional operation, the "nonpower" side of the gear's teeth is not finished. Therefore, this side is rougher and does not provide the same tolerance as the finished "power" side.

Note that it has become standard practice in some plants to reverse the pinion or bullgear in an effort to extend the gear set's useful life. While this practice permits longer operation times, the torsional power generated by a reversed gear set is not as uniform and consistent as when the gears are properly installed.

Gear overload is another leading cause of failure. In some instances, the overload is constant, which is an indication that the gearbox is not suitable for the application. In other cases, the overload is intermittent and occurs only when the speed changes or specific production demands cause a momentary spike in the torsional load requirement of the gearbox.

Misalignment, both real and induced, is another primary root cause of gear failure. The only way to assure that gears are properly aligned is to hard blue the gears immediately following installation. After the gears have run for a short time, their wear pattern should be visually inspected. If the pattern does not conform to vendor's specifications, the alignment should be adjusted.

Table 26–1 Common Failure Modes of Gearboxes and Gear Sets

THE CAUSES	Gear Failures	Variations In Torsional Power	Insufficient Power Output	Overheated Bearings	Short Bearing Life	Overload on Driver	High Vibration	High Noise Levels	Motor Trips
Bent Shaft				●	●	●	●		
Broken or Loose Bolts or Setscrews				●			●		
Damaged Motor						●	●		●
Eliptical Gears		●	●			●	●		
Exceeds Motor's Brake Horsepower Rating			●			●			
Excessive or Too Little Backlash	●	●							
Excessive Torsional Loading	●	●	●	●	●	●			●
Foreign Object In Gearbox	●						●	●	●
Gear Set Not Suitable for Application	●		●			●	●		
Gears Mounted Backward on Shafts			●				●	●	
Incorrect Center-to-Center Distance Between Shafts							●	●	
Incorrect Direction of Rotation			●			●	●		
Lack of or Improper Lubrication	●	●		●	●		●	●	●
Misalignment of Gears or Gearbox	●	●		●	●		●	●	
Overload	●		●	●	●	●			
Process Induced Misalignment	●	●		●	●				
Unstable Foundation		●		●			●	●	
Water or Chemicals in Gearbox	●								
Worn Bearings							●	●	
Worn Coupling							●		

Source: Integrated Systems, Inc.

Poor maintenance practices are the primary source of real misalignment problems. Proper alignment of gear sets, especially large ones, is no easy task. Gearbox manufacturers do not provide an easy, positive means to assure that shafts are parallel and that the proper center-to-center distance is maintained.

Induced misalignment also is a common problem with gear drives. Most gearboxes are used to drive other system components, such as bridle or process rolls. If misalignment is present in the driven members (either real or process induced), it also will directly affect the gears. The change in load zone caused by the misaligned driven component will induce misalignment in the gear set. The effect is identical to real misalignment within the gearbox or between the gearbox and mated (i.e., driver and driven) components.

Visual inspection of gears provides a positive means to isolate the potential root cause of gear damage or failures. The wear pattern or deformation of gear teeth provide clues as to the most likely forcing function or cause. The following sections discuss the clues that can be obtained from visual inspection.

NORMAL WEAR

Figure 26–1 illustrates a gear that has a normal wear pattern. Note that the entire surface of each tooth is uniformly smooth above and below the pitch line.

ABNORMAL WEAR

Figures 26–2 through 26–4 illustrate common abnormal wear patterns found in gear sets. Each of these wear patterns suggests one or more potential failure modes for the gearbox.

Abrasion

Abrasion creates unique wear patterns on the teeth. The pattern varies, depending on the type of abrasion and its specific forcing function. Figure 26–2 illustrates severe abrasive wear caused by particulates in the lubricating oil. Note the score marks that run from the root to the tip of the gear teeth.

Figure 26–1 Normal wear pattern.

Figure 26–2 Wear pattern caused by abrasives in lubricating oil.

Chemical Attack or Corrosion

Water and other foreign substances in the lubricating oil supply also cause gear degradation and premature failure. Figure 26–3 illustrates a typical wear pattern on gears caused by this failure mode.

Figure 26–3 Pattern caused by corrosive attack on gear teeth.

Figure 26–4 Pitting caused by gear overloading.

Overloading

The wear patterns generated by excessive gear loading vary, but all share similar components. Figure 26–4 illustrates pitting caused by excessive torsional loading. The pits are created by the implosion of lubricating oil. Other wear patterns, such as spalling and burning, also can help identify specific forcing functions or root causes of gear failure.

27

STEAM TRAPS

Most of the failure modes that affect steam traps can be attributed to variations in operating parameters or improper maintenance. Table 27–1 lists the more common causes of steam trap failures.

Operation outside the trap's design envelope results in loss of efficiency and may cause premature failure. In many cases, changes in the condensate load, steam pressure or temperature, and other related parameters are the root cause of poor performance or reliability problems. Careful attention should be given to the actual versus design system parameters. Such deviations often are the root causes of problems under investigation.

Poor maintenance practices or the lack of a regular inspection program may be the primary source of steam trap problems. It is important for steam traps to be routinely inspected and repaired to assure proper operation.

Table 27–1 Common Failure Modes of Steam Traps

THE CAUSES	Trap Will Not Discharge	Will Not Shut-off	Continuously Blows Steam	Capacity Suddenly Falls Off	Condensate Will Not Drain	Not Enough Steam Heat	Traps Freeze In Winter	Back Flow In Return Line
Back-pressure Too High				●				
Boiler Foaming or Priming		●				●		
Boiler Gauge Reads Low	●							
Bypass Open or Leaking	●		●					
Condensate Load Greater Than Design		●						
Condensate Short-Circuits					●			
Defective Thermostatic Elements						●		
Dirt or Scale In Trap			●		●			
Discharge Line Has Long Horizontal Runs							●	
Flashing in Return Main				●				●
High-Pressure Traps Discharge Into Low-pressure Return								●
Incorrect Fittings or Connectors				●				●
Internal Parts of Trap Broken or Damaged	●	●	●		●			
Internal Parts of Trap Plugged	●				●			
Kettles or Other Units Increasing Condensate Load		●						
Leaky Steam Coils		●						
No Cooling Leg Ahead of Thermostatic Trap						●		●
Open Bypass or Vent In Return Line				●				
Pressure Regulator Out of Order	●							
Process Load Greater Than Design		●						
Plugged Return Lines				●				
Plugged Strainer, Valve, or Fitting Ahead of Trap	●							
Scored or Out-of-Round Valve Seat in Trap						●		
Steam Pressure Too High	●							
System is Air-bound					●			
Trap and Piping Not Insulated							●	
Trap Below Return Main				●				●
Trap Blowing Steam Into Return				●				
Trap Inlet Pressure Too Low				●	●			
Trap Too Small for Load		●						

Source: Integrated Systems, Inc.

28

INVERTERS

Table 28–1 lists the common symptoms and causes of inverter problems. Most of these problems can be attributed to improper selection for a particular application. Others are caused by improper operation.

When evaluating inverter problems, careful attention should be given to recommendations found in the vendor's operations and maintenance manual. These recommendations often are extremely helpful in isolating the true root cause of a problem.

Table 28–1 Common Failure Modes of Inverters

THE CAUSES	Main Circuit Undervoltage	Control Circuit Undervoltage	Momentary Power Loss	Overcurrent	Ground Fault	Overvoltage	Load Short-Circuit	Heat-sink Overheat	Motor/Inverter Overload	Frequent Speed Deviations
Accel/Decel Time Too Short				•						•
Acceleration Rate Too High									•	•
Ambient Temperature Too High								•		
Control Power Source Too Low			•							
Cooling Fan Failure or Improper Operation								•		
Deceleration Time Too Short						•				•
Excessive Braking Required						•				
Improper or Damaged Power Supply Wiring	•	•								
Improper or Damaged Wiring in Inverter-Motor					•					
Incorrect Line Voltage	•	•				•				
Main Circuit DC Voltage Too Low			•							
Motor Coil Resistance Too Low				•			•			
Motor Insulation Damage				•	•					
Pre-charge Contactor Open			•							
Process Load Exceeds Motor Rating									•	•
Process Load Variations Exceed System Capabilities										•

Source: Integrated Systems, Inc.

29

CONTROL VALVES

Although there are limited common control valve failure modes, the dominant problems are usually related to leakage, speed of operation, or complete valve failure. Table 29–1 lists the more common causes of these failures.

Special attention should be given to the valve actuator when conducting a root cause failure analysis. Many of the problems associated with both process and fluid-power control valves really are actuator problems.

In particular, remotely controlled valves that use pneumatic, hydraulic, or electrical actuators are subject to actuator failure. In many cases, these failures are the reason a valve fails to properly open, close, or seal. Even with manually controlled valves, the true root cause can be traced to an actuator problem. For example, when a manually operated process-control valve is jammed open or closed, it may cause failure of the valve mechanism. This overtorquing of the valve's sealing device may cause damage or failure of the seal, or it may freeze the valve stem. Either failure mode results in total valve failure.

Table 29–1 Common Failure Modes of Control Valves

	THE CAUSES	Valve Fails to Open	Valve Fails to Close	Leakage Through Valve	Leakage Around Stem	Excessive Pressure Drop	Opens/Closes Too Fast	Open/Closes Too Slow
Manually Actuated	Dirt/Debris Trapped In Valve Seat		●	●				
	Excessive Wear		●	●				
	Galling	●	●					
	Line Pressure Too High	●	●	●	●	●		
	Mechanical Damage	●	●					
	Not Packed Properly				●			
	Packed Box Too Loose				●			
	Packing Too Tight	●	●					
	Threads/Lever Damaged	●	●					
	Valve Stem Bound	●	●					
	Valve Undersized					●		●
Pilot Actuated	Dirt/Debris Trapped In Valve Seat	●	●	●				
	Galling	●	●					
	Mechanical Damage (Seals, Seat)	●	●	●				
	Pilot Port Blocked/Plugged	●	●	●				
	Pilot Pressure Too High		●				●	
	Pilot Pressure Too Low	●		●				●
Solenoid Actuated	Corrosion	●	●	●				
	Dirt/Debris Trapped In Valve Seat	●	●	●				
	Galling	●	●					
	Line Pressure Too High	●	●	●	●			●
	Mechanical Damage	●	●	●				
	Solenoid Failure	●	●					
	Solenoid Wiring Defective	●	●					
	Wrong Type of Valve (N-O, N-C)	●	●					

Source: Integrated Systems, Inc.

30

SEALS AND PACKING

Failure modes that affect shaft seals normally are limited to excessive leakage and premature failure of the mechanical seal or packing. Table 30–1 lists the common failure modes for both mechanical seals and packed boxes. As the table indicates, most of these failure modes can be attributed directly to misapplication, improper installation, or poor maintenance practices.

MECHANICAL SEALS

By design, mechanical seals are the weakest link in a machine train. If there is any misalignment or eccentric shaft rotation, the probability of a mechanical seal failure is extremely high. Most seal tolerances are limited to no more than 0.002 in. of total shaft deflection or misalignment. Any deviation outside this limited range will cause catastrophic seal failure.

Misalignment

Physical misalignment of a shaft will either cause seal damage and permit some leakage through the seal or it will result in total seal failure. Therefore, it is imperative that good alignment practices be followed for all shafts that have an installed mechanical seal.

Process- and machine-induced shaft instability also creates seal problems. Primary causes for this failure mode include aerodynamic or hydraulic instability, critical speeds, mechanical imbalance, process load changes, or radical speed changes. These can cause the shaft to deviate from its true centerline enough to result in seal damage.

Table 30–1 Common Failure Modes of Packing and Mechanical Seals

			THE PROBLEM							
		THE CAUSES	Excessive Leakage	Continuous Stream of Liquid	No Leakage	Shaft Hard to Turn	Shaft Damage Under Packing	Frequent Replacement Required	Bellows Spring Failure	Seal Face Failure
Packed Box	**Nonrotating**	Cut Ends of Packing Not Staggered	●	●				●		
		Line Pressure Too High	●							
		Not Packed Properly				●	●	●		
		Packed Box Too Loose	●	●						
		Packing Gland Too Loose	●	●						
		Packing Gland Too Tight	●	●		●	●	●		
	Rotating	Cut End of Packing Not Staggered		●						
		Line Pressure Too High	●							
		Mechanical Damage (Seals, Seat)	●	●	●			●		
		Incompatible Packing	●	●			●			
		Packing Gland Too Loose	●							
		Packing Gland Too Tight				●	●	●		
Mechanical Seal	**Internal Flush**	Flush Flow/Pressure Too Low							●	●
		Flush Pressure Too High	●	●					●	●
		Improperly Installed	●						●	●
		Induced Misalignment	●							
		Internal Flush Line Plugged							●	●
		Line Pressure Too High							●	●
		Physical Shaft Misalignment	●							
		Seal Not Compatible With Application	●							
	External Flush	Contamination In Flush Liquid	●							●
		External Flush Line Plugged							●	●
		Flush Flow/Pressure Too Low							●	●
		Flush Pressure Too High	●	●					●	●
		Improperly Installed	●							●
		Induced Misalignment	●						●	●
		Line Pressure Too High							●	●
		Physical Shaft Misalignment	●						●	●
		Seal Not Compatible with Application	●							●

Source: Integrated Systems, Inc.

Chemical Attack

Chemical attack (i.e., corrosion or chemical reaction with the liquid being sealed) is another primary source of mechanical seal problems. Generally, two primary factors cause chemical attack: misapplication or improper flushing of the seal.

Misapplication

Little attention generally is given to the selection of mechanical seals. Most plants rely on the vendor to provide a seal that is compatible with the application. Too often there is a serious breakdown in communications between the end user and the vendor on this subject. Either the procurement specification does not provide the vendor with appropriate information or the vendor does not offer the option of custom ordering the seals. Regardless of the reason, mechanical seals often are improperly selected and used in inappropriate applications.

Seal Flushing

When installed in corrosive chemical applications, mechanical seals must have a clear-water flush system to prevent chemical attack. The flushing system must provide a positive flow of clean liquid to the seal and also an enclosed drain line that removes the flushing liquid. The flow rate and pressure of the flushing liquid will vary, depending on the specific type of seal, but must be enough to assure complete, continuous flushing.

PACKED BOXES

Packing is used to seal shafts in a variety of applications. In equipment where the shaft is not continuously rotating (e.g., valves), packed boxes can be used successfully with no leakage around the shaft. In rotating applications, such as pump shafts, the application must be able to tolerate some leakage around the shaft.

Nonrotating Applications

In nonrotating applications, packing can be installed tight enough to prevent leakage around the shaft. As long as the packing is properly installed and the stuffing-box gland is properly tightened, there is very little probability of seal failure. This type of application requires periodic maintenance to ensure that the stuffing-box gland is properly tightened or that the packing is replaced when required.

Rotating Applications

In applications where a shaft continuously rotates, packing cannot be tight enough to prevent leakage. In fact, some leakage is required to provide both flushing and cooling of the packing. Properly installed and maintained packed boxes should not fail or contribute to equipment reliability problems. Proper installation is relatively easy and routine maintenance is limited to periodic tightening of the stuffing-box gland.

31

OTHERS

This chapter provides some general troubleshooting guides that may assist with a root cause failure analysis.

The wear pattern or surface finish of parts may provide clues that will help the investigator resolve a problem. Table 31–1 provides a guide to the meaning of these patterns.

Bolting practices and bolted joints may be a contributor to equipment reliability problems. Table 31–2 provides troubleshooting guidelines for this common machine-train component.

Table 31–1   Common Failure Modes of Wear Part Surfaces

	Description of Worn Surface	Staining	Pitting	Spalling	Cavitation	Erosion	Abrasive Wear	Fretting	Polishing	Scratching	Scuffing	Gouging	Scoring	Grooving	Galling	Exfoliation	Melting
Micro-Smooth	Melted (Bubbles.Wavy)																•
	Progressive loss and reformation of surface films by fine abrasion and/or tractive stresses, mutually imposed by adhesive or viscous interaction							•	•		•			•	•		
	Very fine abrasion with loss of substrate in addition to loss of surface film					•			•								
Micro-Rough	Abrasion by medium-coarse particles						•			•		•					
	Due to tractive stresses resulting from adhesion														•		
	Micro-pitting by fatigue		•		•												
Macro-Smooth	Caused by abrasive held on or between solid backing						•			•	•						
Macro-rough	Advanced stages of micro-roughening, where little unaffected surface remains between pits		•					•									
	Abrasion by coarse particles, including carbide and other hard inclusions in the sliding materials												•	•			
	Abrasion by fine particles in turbulent fluid, producing scallops, waves, etc.					•											
	Local fatigue failure resulting in pits or depressions caused by repeated rolling-contact stress and high-friction sliding or impact by hard particles, as in erosion		•	•	•	•										•	
	Severe adhesion										•				•		
Dull	Thick films resulting from aggressive environments, including high temperature due to corrosion	•						•									
Shiny	Very thin, or no, surface film of oxide, hydroxide, sulfide, chloride, or other							•	•	•				•		•	

Source: Integrated Systems, Inc.

Table 31–2 Common Failure Modes of Bolted Joints

		Fracture Under Static Load	Fatigue Failure	Vibration Loosening	Joint Leakage	Bolt Failure
	THE CAUSES					
Design and Manufacturing	Bolt/Joint Stiffness Ratio		●		●	●
	Damping in Joint			●		
	Direction of Bolt Axis Relative to Vibration Axis			●		
	Fillet Size and Shape		●			
	Galling	●				
	Improper Heat Treatment	●				
	Nut Dilation	●				
	Parts Finish		●			
	Poor Fit	●				●
	Radius of Thread Roots		●			
	Relaxation Effects		●	●		
	Thread Runout		●			
	Tool Marks		●			
Assembly Practices	Bolt-up Procedure				●	●
	Condition of Gaskets				●	
	Condition of Joint Surfaces				●	
	Improper Preload			●	●	●
	Thread Lubrication	●	●	●		
	Type of Tool Used	●				
Operating Conditions	Corrosion	●	●	●	●	
	Magnitude of Load Excursions	●	●			
	Temperature Cycling				●	

Source: Integrated Systems, Inc.

LIST OF ABBREVIATIONS

AC	Alternating current
BEP	Best efficiency point
BHP	Brake horsepower
cfm	Cubic feet per minute
CA	Cellulose acetate
CERCLA	Comprehensive Environmental Response, Compensation, and Liability Act
CFR	Code of Federal Regulations
CWA	Clean Water Act
DC	Direct current
DOT	Department of Transportation
EPA	Environmental Protection Act
EPCRA	Emergency Planning and Community Right-to-Know Act
FC	Fan capacity
FMEA	Failure mode and effects analysis
fpm	Feet per minute
gpm	Gallons per minute
HAZWOPER	Hazardous Waste Operations and Emergency Response

HMTA	Hazardous Materials Transportation Act
hp	Horsepower
I.D.	Inside diameter
LEPC	Local Emergency Planning Committee
LPG	Liquified petroleum gas
ME	Mechanical efficiency
MSDS	Material safety data sheets
NPSH	Net positive suction head
NPSHA	Net positive suction head, available
NPSHR	Net positive suction head, required
NPSHA	Actual net positive suction head
NRC	National Response Center
O.D.	Outside diameter
O&M	Operating and maintenance
OSHA	Occupational Safety and Health Act (or Agency)
OV	Outlet velocity
PL	Packing length
psi	Pounds per square inch
psia	Pounds per square inch, absolute
psid	Pounds per square inch, differential
psig	Pounds per square inch, gauge
PPE	Personal protection equipment
PSM	Process safety management
PV	Pressure-volume
PVT	Pressure-volume-temperature
RCFA	Root cause failure analysis
RCRA	Resource Conservation and Recovery Act
rpm	Rotating speed
rpm2	Rotating speed, squared
rpm3	Rotating speed, cubed

SARA	Superfund Amendments and Reauthorization
SE	Static efficiency
SERC	State Emergency Response Committee
SMP	Standard maintenance procedures
SOP	Standard operating procedures
SP	Static pressure
TDH	Total dynamic head
TIR	Total indicated runout
TP	Total pressure
TS	Tip speed
TSCA	Toxic Substances Control Act
TSH	Total system head
V/Hz	Volts per hertz
VP	Velocity pressure
X, 2X, . . .	One times, two times, . . .

GLOSSARY

acoustic emissions

Testing technique that can be used to determine the structural integrity of objects (e.g., pressure vessels).

actuate

To put into motion or mechanical action, as by an actuator.

actuator

A mechanism to activate process-control equipment by use of pneumatic, hydraulic, or electronic signals.

aeration

Exposing a liquid to air through bubbling.

aerodynamic instability

An unstable state caused by oscillations of a structure that are generated by spontaneous and more or less periodic fluctuations in the flow, particularly in the wake of the structure.

agitator

A device for keeping liquids and solids in liquids in motion by mixing, stirring, or shaking.

ambient condition

Surrounding condition, especially pertaining to temperature, pressure, humidity, and the like.

amplitude

The magnitude or size of a quantity such as velocity, displacement, or acceleration, measured by a vibration analyzer in conjunction with a displacement probe, velocity transducer, or accelerometer.

analytical

Describes the process of breaking up a whole into its parts to find out their nature.

angstrom	A unit of length equal to 100 millionth (10^{-8}) of a centimeter.
annotation	An explanatory note.
assumption	Describes unconfirmed or perceived factors that may have contributed to an event; used when confirmed qualifiers are unavailable.
axial	Of, on, around, or along an axis (straight line about which an object rotates) or center of rotation.
backlash	Amount by which a gear tooth space exceeds the thickness of the engaging tooth on the operating pitch circles.
back pressure	Pressure due to a force that is operating in a direction opposite to that being considered, such as that of a fluid flow.
baghouse	The large chamber or room for holding bag filters used to filter gas streams.
bearing	A machine element that supports a part, such as a shaft, that rotates, slides, or oscillates in or on it. Three major types are radial or journal, thrust, and guide bearings.
blower	A fan that operates where the resistance to gas flow is predominantly downstream of the fan.
boundary condition	Any condition that marks or describes a limit or a bound.
brake horsepower	The horsepower developed by an engine as measured by the force applied to a friction brake or by an absorption dynamometer applied to the shaft or flywheel.
catastrophic failure	A sudden failure without warning, as opposed to degradation failure, that generally prevents the satisfactory performance of an entire assembly or system.
cavitation	Formation of gas- or vapor-filled cavities within liquids.
centrifugal force	An outward pseudo-force in a reference frame that is rotating with respect to an inertial reference frame. The pseudo-force is equal and opposite to the centripetal force that acts on a particle stationary in the rotating frame.

centripetal force	The radial force required to keep a particle or object moving in a circular path, which can be shown to be directed toward the center of the circle.
chronic	Of continued duration.
chronology	The arrangement of data in order of time of appearance.
classification	Sorting or categorizing by established criteria.
compressor	A machine used to increase the pressure of a gas or vapor.
concentricity	When the smaller of two circular, cylindrical, or spherical shapes is centered within the larger one.
condensate	The liquid product that forms from condensable gases or vapors when they are subjected to appropriately altered conditions of temperature or pressure.
control valve	A valve that controls pressure, volume, or flow direction in a fluid transmission system.
conveyor	Any materials-handling machine designed to move individual articles, such as solids or free-flowing bulk materials, over a horizontal, inclined, declined, or vertical path of travel with continuous motion.
corrosion	Gradual destruction of a metal or alloy due to chemical processes such as oxidation or the action of a chemical agent.
critical speed	The angular speed at which a rotor becomes dynamically unstable with large lateral amplitudes due to resonance with the natural frequencies of lateral vibration of the rotor.
current in-rush	Initial current surge that occurs on startup of a piece of equipment, such as a motor.
cyclone separator	A funnel-shaped device for removing particles from air or other fluids, steam from water, water from steam, and in certain applications to separate particles into two or more size classes.
deadheading	Pumping against a closed discharge valve.

deductive	The process of deriving a statement from certain assumed statements by applying the rules of logic.
density	The mass of a substance per unit of volume.
detent	A catch or lever in a mechanism that locks the movement of a part, especially in escapement mechanisms or ratchet devices that permit motion in one direction slowly. In a valve, a detent is used to hold the valve in its last position.
deviation	The difference between the actual value of a controlled variable and the desired or average value.
dial indicator	Meter or gauge with a calibrated circular face and a pivoted pointer to give readings.
digital	Data having discrete values, as contrasted with continuous analog data. Digital data recorders convert continuous electrical analog signals into numbered (digital) values and record these values.
disseminate	Distribute widely.
divergent	Moving in different directions.
ductwork	A system of enclosed runways.
duplex cylinder	Cylinders having overlapping bores.
dynamic pressure	The pressure that a moving fluid would have if it were brought to rest by isentropic flow against a pressure gradient; also known as *impact pressure*, *stagnation pressure*, and *total pressure*.
dynamic response	The response, measured over time, of a component or system to a forcing function.
dynamics, operating	Deals with the motion of a system under the influence of forces, especially those forces that originate outside the system under consideration.
eccentric	The condition when a disk or wheel has its axis of revolution displaced from the center so that it is capable of imparting reciprocating motion.

eccentricity	The distance of the geometric center of a revolving body from the axis of rotation.
efficiency	Ratio of useful energy provided by a dynamic system to the energy supplied to it during a specific period of operation.
end play	Lateral or axial shaft movement.
entrainment	Entrapment of tiny air bubbles in a fluid.
envelope, operating	Specifies the physical requirements, dimensions, and type of coupling needed in a specific application. Envelope information includes shaft sizes, orientation of shafts, required horsepower, full-range of operating torque, speed ramp rates, and any other data that directly or indirectly affect the coupling.
event	A specific change that occurs at a specified time.
expedite	To facilitate or speed up a progress.
extrapolation	Making an estimate of an unknown quantity based on, but outside the range of, known data.
facilitator	Someone or something that makes it easier to accomplish a task.
fan	A device, usually consisting of a rotating paddle wheel or an airscrew, with or without a casing, for producing currents to circulate, exhaust, or deliver large volumes of air or gas.
fault event	An abnormal system state.
flange	Projecting rim of a mechanical part.
flash point	The lowest temperature at which vapors from a volatile liquid will ignite momentarily on the application of a small flame under specified conditions.
fluidizer	Identical to single-stage, screw-type compressors or blowers.
forcing function	(1) A cause of an event. (2) The cause of each discrete frequency component in a machine train's vibration signature.

fretting corrosion	Surface damage usually in an air environment between two surfaces, one or both of which are metals, in close contact under pressure and subject to a slight relative motion; also known as *chafing corrosion*.
friction	Resistance to sliding, a property of the interface between two solid bodies in contact. Friction wastefully consumes energy and wear changes dimensions.
full-fluid film	The ideal state of lubrication where the film remains thick and prevents contact between the surface peaks, which is apparent in a microscopic examination of two surfaces; also referred to as *thick-film lubrication*.
fuse	An expendable device for opening an electric circuit when the current therein becomes excessive. The device contains a section of conductor that melts when the current through it exceeds a rated value for a definite period of time.
gear	Disk or wheel with teeth around its periphery, either on the inside edge (i.e., internal gear) or on the outside edge (i.e., external gear), used to provide a positive means of power transmission.
gearbox	The gearing system by which power is transmitted from the source to the rotating shaft or axle.
gland	A device for preventing leakage at a machine joint, as where a shaft emerges from a vessel containing a pressurized fluid; also referred to as a *packed-stuffing box*.
hard blue	A temporary coating used on mating parts to evaluate fit and wear patterns.
head	The height of a column of fluid necessary to develop a specific pressure.
hydraulic	Operated or effected by the action of water or other low-viscosity fluid.
hydraulic curve	A plot with total dynamic head as the vertical axis and discharge volume or flow as the horizontal axis.

hydraulic hammer	An instantaneous shock load created by rapidly closing a valve. A radical surge of back pressure is generated that can tear a pump or other equipment from its piping and foundation.
impeller	Rotating member or rotor of a turbine, blower, fan, axial or centrifugal pump, or mixing apparatus that physically moves a fluid; also known as a *rotor*.
incident	Something that happens, an event.
incremental cost	The difference between the costs of two alternative procedures or operating conditions.
indices	Indicators.
inductive	The process of reasoning from particular facts to a general conclusion.
inertia	The property of an object by which it remains at rest or in uniform motion in the same straight line unless acted on by some external force.
inverter	A device for converting direct current into alternating current; it may be electromechanical or electronic; also known as a *DC-to-AC converter*.
involute tooth profile	The profile most commonly used for gear teeth. It does not provide as good a rolling action as the cycloidal profile and, as a result, some sliding between teeth occurs. It is less sensitive to shaft alignment and gear spacing than the cycloidal profile.
kinetic energy	Energy associated with motion.
laminar	Viscous streamline flow without turbulence.
lateral	Of, at, on, or toward the side.
lift, dynamic	The change in vertical distance that occurs when a system is in operation.
lift, static	The vertical distance present regardless of whether a system is operating or not.
lift, suction	*See* Suction lift.
lift, vertical	The elevation change, measured from centerline to centerline.

longitudinal	Pertaining to the lengthwise dimension.
lubricity	The ability of a material to lubricate.
machine train	A series of machines containing both driver and driven components.
mandrel	A shaft inserted through a hole in a component for support.
manifold	Piping system that either gathers multiple-line fluid inputs into a single intake chamber (intake manifold) or divides a single fluid supply into several outlet streams (distribution manifold).
mechanical imbalance	A condition that can result from a mechanical or force imbalance. Mechanical imbalance occurs when there is more weight on one side of a centerline of a rotor than on the other. Force imbalance can result when there is an imbalance of the centripetal forces generated by rotation or when there is an imbalance between the lift generated by the rotor and gravity.
mechanism	Any physical or mental process by which a result is produced.
meggering	Testing technique that can be used to measure the cause-and-effect relationship between variables in electric motors.
meshing	Engagement or working contact of teeth of gears or of a gear and a rack.
methodology	A system of methods.
mixer	A device used to cause intermingling of different materials (liquid, gas, solid) to produce a homogeneous mixture.
multichannel analysis	Analysis based on data acquired simultaneously from multiple measurement points.
noise, white	Random noise that has a constant energy per unit bandwidth at every frequency in the range of interest.
normal event	An event that is expected to occur.
oblique	Having a slanted direction or position at an angle that is neither a right angle nor a multiple of a right angle.

operating dynamics	Deals with the motion of a system under the influence of forces, especially those that originate outside the system under consideration.
operating envelope	The physical requirements and dimensions of a specific application.
orifice	An aperture or hole.
oscillator	An electronic circuit that converts energy from a direct-current source to a periodically varying electric output.
parameter	A quantity that is constant under a given set of conditions but may be different under other conditions.
peripheral	Pertaining to or located at or near the surface of a part.
pinion	The smaller of a pair of gear wheels or the smallest wheel of a gear train.
piston	A sliding metal cylinder that reciprocates in a tubular housing, either moving against or moved by fluid pressure.
pitch	Distance between similar, equally spaced gear tooth surfaces along a given line or curve.
pitch, diametrical	A gear tooth design factor that is a measure of tooth size in the English system; expressed as the ratio of the number of teeth to the diameter of the pitch circle measured in inches.
pneumatic	Pertaining to or operated by air or other gas.
positive displacement compressor	A device that confines successive volumes of fluid within a closed space in which the pressure of the fluid is increased as the volume of the closed space is decreased.
positive displacement pump	A device in which a measured quantity of liquid is entrapped in a space, its pressure is raised, and then it is delivered.
predictive maintenance	The practice of using actual operating conditions of plant equipment and systems to optimize total plant operation. It relies on direct equipment monitoring to determine the actual mean-time-to-failure or loss of efficiency for each machine train and system in a plant. This technique is used in place of traditional run-to-failure programs.

primary event	A direct, although possibly not obvious, contributor to the top event in a fault-tree analysis.
prime	Add fluid to a pump to enable it to begin pumping.
probability	The probability of an event is the ratio of the number of times it occurs to the large number of trials that take place.
proportional	The comparative relation between things in size, amount, etc. (i.e., ratio).
pulsation damper	A device installed in a fluid piping system (gas or liquid) to eliminate or even out the fluid-flow pulsations caused by reciprocating compressors, pumps, and such.
pump	A machine that draws a fluid into itself through an entrance port and forces the fluid out through an exhaust port.
qualifier	Statement that provides confirmed background or support data needed to accurately define an event or forcing function.
quantifiable	Describes something that can be expressed as a quantity or a number.
quantitative	Having to do with quantity.
race (inner and outer)	Either of the concentric pair of steel rings of a ball or roller bearing.
radial	Radiating from or converging to a common center.
radius	A line segment joining the center and a point on the outside of a circle or sphere.
real-time	Responses to an event are essentially simultaneous with the event itself.
real-time data	Responses to an event that essentially are simultaneous with the event itself.
reciprocation	The action of moving back and forth alternately.
recur	To occur again or at intervals.
reducer	*See* Gearbox.

regulatory compliance	Regulatory-compliance issues associated with work-related incidents generally fall into the domains of safety, which is regulated by OSHA, and the environment, which is regulated by the EPA and state and local governments.
reliability engineering	Discipline dealing with the probability that a component part, equipment, or system will satisfactorily perform its intended function under given circumstances, such as environmental conditions, limitations as to operating time, and frequency and thoroughness of maintenance for a specified period of time.
resonance	A large amplitude vibration in a mechanical system caused by a small periodic stimulus of the same or nearly the same period as the natural vibration period of the system. Higher levels of input energy can result in catastrophic, near instantaneous failure of the machine or structure.
roll	A cylinder mounted in bearings used for such functions as shaping, crushing, moving, or printing work passing by it.
roller spin	The speed at which a bearing's roller turns in its races.
root cause	The true source of a problem or event.
rotor	Any rotating part of a machine.
runout	Axial or radial looseness. A measure of shaft wobble caused by being off center.
saturation point or pressure	The pressure of a thermodynamic system, at a given temperature, where the vapor of a substance is in equilibrium with that substance's pure liquid or solid phase.
schematic diagram	A presentation of the element-by-element relationship of all parts of a system.
score marks	Parallel scratches, lines, or grooves.
seal	A means of preventing migration of fluids, gases, or particles across a joint or opening.
seat	The fixed, pressure-containing portion of a valve that comes into contact with the moving portions of that valve.

shaft encoder	A device that senses the true running speed of the shaft and sends a signal back to the speed-control mechanism; also referred to as a *resolver.*
shroud	A protective covering, usually of metal plate or sheet.
single-channel data	Acquired in series or one channel at a time.
sinusoidal function	The real or complex function $\sin(u)$ or any function with analogous continuous periodic behavior.
slip	For a pump, it is the rate at which liquid leaks from the discharge side to the suction side. For a conveyor, it is the amount of material that bypasses the system.
solenoid	An electrically energized coil of insulated wire that produces a magnetic field within the coil.
specific gravity	The ratio of a material's density to the density of some standard material (i.e., water at 60°F or air at standard temperature and pressure); also known as *relative density.*
stagnant	Motionless.
static head	Pressure of a fluid due to the head of fluid above some reference point.
static pressure	The pressure exerted in a nonoperating system.
steam trap	A device that separates and removes condensate automatically from steam lines.
suction lift	The head, in feet, that a pump must provide on the inlet side to raise the liquid from the supply well to the level of the pump; also known as the *suction head.*
synchronize	Make to occur at the same time.
three-phase alternating current	Alternating current delivered through three wires, with each wire serving as the return for the other two and with the three current components differing in phase successively by one-third cycle, or 120 electrical degrees.
throttle	A choking device to regulate flow of a fluid.

thrust	Axial forces and vibration created by the mechanical or dynamic operation of machine trains or process systems.
top event	A system failure in a fault-tree analysis; it can be either a broad, all-encompassing failure or a failure of a particular component.
torque	A moment/force couple applied to a rotor such as a shaft to sustain acceleration/load requirements. A twisting load imparted to shafts as the result of induced loads/speeds.
torsion	A straining action produced by couples that act normal to the axis of a member, identified by a twisting deformation.
total system head	The total pressure required to overcome all downstream resistance at a given flow.
transformer	An electrical component required when a different input or output voltage is needed for proper operation of equipment.
transient	An event such as an impact or change in speed or process parameter that varies in time. A temporary phenomenon occurring in a system prior to reaching a steady-state condition.
turbulent flow	Motion of fluids in which local velocities and pressures fluctuate irregularly, in a random manner.
ultrasonic analysis	Testing technique that can be used to measure the cause-and-effect relationship between variables. It uses principles similar to those of vibration analysis to monitor the noise generated by plant machinery or systems to determine their actual operating condition. Ultrasonics is used to monitor the higher frequencies (i.e., ultrasound) that range between 20,000 Hz and 100 kHz.
unloading	Depressuring or emptying a process unit.
vane	Casting on the backside of the impeller that reduces the pressure acting on the fluid behind the impeller.
variable	Data item or symbol that can assume any of a set of values.

velocity head	The square of the speed of flow of a fluid divided by twice the acceleration of gravity. It is equal to the static pressure head corresponding to a pressure equal to the kinetic energy of the fluid per unit volume.
vibration analysis	Predictive-maintenance technique that monitors the noise or vibrations generated by plant machinery or systems to determine their actual operating condition. The normal monitoring range for vibration analysis is from less than 1 to 20,000 Hz.
vibration profile	Refers to either time-domain (also called *time trace* or *waveform*) or frequency-domain vibration curves.
vibration signature	Frequency-domain vibration curve, which is amplitude (displacement, velocity, or acceleration) versus frequency. Frequency-domain data are obtained by converting time-domain data using a mathematical technique referred to as *fast Fourier transform*. Time-domain data are plotted as amplitude versus time.
viscosity	The resistance that a gaseous or liquid system offers to flow when subjected to a shear stress; also known as *flow resistance* or *internal friction*.
viscous fluid	A fluid whose viscosity is sufficiently large to make the viscous forces a significant part of the total force field in the fluid.
volute	Spiral part of a pump casing that accepts the fluid from the impeller and carries it to the outlet.
vortex	Flow exhibiting rotary motion as in an eddy or whirlpool.

REFERENCES

Baumeister, Theodore, Ed. *Marks' Standard Handbook for Mechanical Engineers*, 8th ed. New York: McGraw-Hill, 1978.

Bearings Inc. Catalogue. Bearings Inc.

Gibbs, Charles W. *Compressed Air and Gas Data*. Ingersoll-Rand Company, 1971.

Higgins, Lindley R., and R. Keith Mobley. *Maintenance Engineering Handbook*. New York: McGraw-Hill, 1995.

Mobley, R. Keith. *Advanced Diagnostics and Analysis*. Plant Performance Group, 1989.

Neale, M. J. *Bearings—A Tribology Handbook*. Society of Automotive Engineers, Inc., Oxford, England: Butterworth–Heinemann Ltd., 1993.

Nelson, Carl A. *Millwrights and Mechanics Guide*. New York: Macmillan, 1986.

95/96 Product Guide. GE Capital Test Equipment Management Services, 1995-1996. 1-800-GE-RENTS.

Perry, Robert H., and Don Green. *Perry's Chemical Engineers' Handbook*, 6th ed. New York: McGraw-Hill, 1984.

Roberts, William L. *Cold Rolling of Steel*. New York and Basel: Marcel Dekker, 1978.

Technical Training Manual-Practical Hydraulics. Vickers, Inc.

Thomas Register of American Manufacturers, 85th ed. New York: Thomas Publishing Co., 1995.

INDEX

Acceleration control, 198
Accidents, 72
Actuators, 217
Aerodynamic instability, 246
Agitators, 147, 264
Application review, 30
Axial fans, 98

Backlash, 184
Backup valves, 214
Backward-curved blades, 99
Baghouses, 153, 266, 267
Ball valves, 202
Best efficiency point, 85
Bevel gears, 177
Bimetallic steam trap, 190
Blowers, 97, 111, 248, 249
Boiler-feed applications, 83
Brake horsepower, 83, 84, 150, 184
Bullgear compressors, 126
Butterfly valves, 204

Capacity loss, 50, 72
Catastrophic failure, 3
Cause-and-effect analysis, 9, 10, 72
Cavitation, 83
Centrifugal compressors, 123, 254
Centrifugal fans, 246, 247
Centrifugal pump, 77, 87, 239
    performance, 79
Chain conveyors, 114, 253
Communications, 44

Compressors, 123, 254
Control valves, 202, 280
Conveyor performance, 113
Conveyors, 112, 251
Corrective actions, 47, 49
Cost avoidance, 53
Cost-benefit analysis, 48, 52, 54
Cut-flight conveyors, 118
Cyclonic Separators, 160, 266, 268

Dampers, 110
Data gathering, 20
Delivery slippage, 50
Department of transportation, 71
Design
    data, 27
    practices, 38, 39
    review, 19
Discharge piping, 88
Dust collectors, 153, 266

Economic performance, 26, 73
Emergency planning, 70
Entrained air or gas, 82, 241
Environmental conditions, 23
Environmental release, 18
Equipment damage, 28
    design, 18, 22, 23, 38
    evaluation, 77
    failures, 28, 72
    nameplate data, 35

Failed components, 25
Failure modes, 6, 7
  and effects analysis, 7, 8, 58
Fan, 97, 246
  capacity, 101
  installation, 105
  laws, 103
  performance, 101
  rating, 101, 106, 107, 108
  speed, 110
Fatigue, 60
Fault-tree analysis, 16
Float-type steam trap, 188
Flow-control devices, 109
Fluidizers, 97, 111
Fluid power, 210
Forward-curved blades, 100
Four-way valves, 213
Friction loss, 80, 88, 113

Gate valves, 203
Gear
  backlash, 184
  efficiency, 183
  ratios, 185
Gearboxes, 171, 269
Globe valves, 205
Ground-fault protection, 200

Hazardous materials transportation, 71
Helical gear pump, 93
Helical gears, 174
Herringbone gear pump, 93
Herringbone gears, 177
Human engineering, 44
Hydraulic curve, 84
Hydraulic slippage, 91
Hypoid gears, 181

Implementation costs, 51
Incident classification, 17, 26
Incident reports, 29, 55
Incoming product, 30, 122
Increased revenue, 53
Inlet piping, 88
Inlet-air conditions, 102
Inline impellers, 78, 79
Inspections, 46
Installation, 20, 24, 113, 121, 129
Interviews, 36

Inverted bucket steam trap, 187
Inverters, 194, 276
Investigator, 57, 69

Life cycle costs, 39
Liquid seal-ring compressors, 133
Lobe pump, 95
Lubrication system, 28

Maintenance manuals, 25
Maintenance records, 30, 32
Maintenance review, 35
Management systems, 45
Measurement devices, 33
Measurements, 33
Mechanical conveyors, 114
Mechanical efficiency, fans, 102
Mechanical seals, 220, 228, 234, 282
Miter gears, 181
Mixers, 147, 264

Natural frequency, 102
Net positive suction head, 80
NPSH, 80

Observations, 27
One-way valves, 211
Operating characteristics, 30, 31, 38
Operating envelope, 28
Operating manuals, 19, 22
Operating performance, 29, 73
Operating procedures, 40, 89, 96
Operations problems, 73
Opposed impellers, 78, 79
OSHA, 58, 62, 65
OSHA reporting requirements, 65, 66, 67
Outlet velocity, 101
Output product, 14

Packing, 220, 234, 282
Perceived causes, 29
Performed work, 25
Personal problems, 60
Physical evidence, 3
Physical impairment, 60
Plate-out in fans, 246
Pneumatic conveyors, 112, 252
Positive displacement compressors, 254, 256
Positive displacement pumps, 90, 243
Predictive maintenance, 14

Problems, 33
Process
  instability, 249
  parameters, 38
  performance, 72
  rolls, 164, 267
Procurement practices, 28, 39
Procurement specifications, 21
Pumps
  in parallel, 81
  in series, 81

Quality control, 46
Questions to ask, 20

Radial blade fans, 100
Reciprocating compressors, 136, 258
Reciprocating pumps, 243, 245
Reducer gear drives, 171, 271
Reduction in unit cost, 52
Regulatory compliance, 14, 62
Repetition of tasks, 61
Report preparation, 54
Reporting problems, 37
Resource conservation and recovery act, 71
Roll face, 166
Roll neck, 166
Root cause, 4, 5, 14
Root cause failure analysis, 19
Rotary compressors, 130, 257
Rotary pumps, 244
Rotating speed, 96

Safety, 10, 11, 12, 13, 23, 27, 58
Screw conveyors, 117
Screw pump, 93
Scrubber rolls, 169
Seals, 220, 280
Sequence-of-events analysis, 31
Setup procedures, 25, 32
Short-circuit protection, 200
Spill hazards, 62
Spill prevention, 63
Spill response, 63
Spiral gears, 180
Spur gear pump, 92

Standard maintenance procedures, 25, 31, 32, 38
Standard operating procedures, 14, 38, 40
Startup procedures, 89
Static efficiency, 101
Static head, 80, 99
Static pressure, 101
Steam traps, 187, 276
Suction volume, 81
Supervision, 42
Symptoms and boundaries, 25

Thermodynamic laws, 127
Thermodynamic steam trap, 189
Thermostatic steam trap, 188
Three-way valves, 212
Tip speed, 101
Total dynamic head, 84
Total system head, 84, 242
Training, 38, 41, 63
Training records, 31
Transient procedures, 28
Two-way valves, 211

Unloading frequency, 136

Valve actuators, 217
Vane pump, 94
Vapor pressure, 80
Vector control, 196
Velocity head, 80, 102
Velocity pressure, 102
Vendor evaluation, 40
Vendor specifications, 33
Vibration analysis, 33
Viscosity, 95, 152
Visual inspection, 35

Wear particles, 38
Work environment, 45
Worker responsibilities, 64
Worker selection, 43
Worker-machine interface, 44
Worm gears, 182

Zerol gears,179